The
Magic
Numbers Of
Dr. MATRIX

The Magic Numbers of

Dr. MATRIX

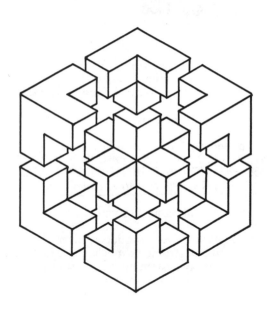

Martin Gardner

Prometheus Books
Buffalo, New York

Published 1985 by
Prometheus Books
700 E. Amherst Street, Buffalo, New York 14215

Portions of this book appeared as
The Numerology of Dr. Matrix and
The Incredible Dr. Matrix

Printed in the United States of America

Library of Congress Catalog Card No. 84-43183
ISBN 0-87975-281-5 cloth
ISBN 0-87975-282-3 paper

For Tom,
my number two son

Contents

Introduction

My friendship with the late Dr. Irving Joshua Matrix and his daughter Iva spanned a period of some twenty years. I first wrote about him in my column on "Mathematical Games" in *Scientific American*, January 1960. With enormous sadness, and echoes of Watson's famous tribute to his friend Sherlock Holmes, my column for September 1980 recorded all I could learn about his untimely death.

This is the third collection of my columns about Dr. Matrix. In 1967 Simon and Schuster brought together the first seven columns in a small book called *The Numerology of Dr. Matrix*. Reverse the digits of 67 and you get the date of *The Incredible Dr. Matrix,* a Scribner's book that contained the previous one. After another lapse of nine years, Prometheus has now allowed me to assemble *all* my Dr. Matrix columns in one volume, from my first meeting with him in Manhattan to his violent end in 1980 on the banks of the Danube.

Over the years, mathematicians and others who have followed my accounts of Dr. Matrix's remarkable predictions, analyses, and play with words and numbers have asked me to provide the doctor's curriculum vitae. I will do the best I can. The information that follows is based almost entirely on what has been disclosed to me over the years by Iva. The reader should understand that, except for a few isolated facts, this information has not been otherwise verified.

Dr. Matrix was born on February 21, 1908, in Kagoshima, on the Japanese island of Kyushu. His father, the Reverend William Miller Bush, was a Seventh-Day Adventist missionary from a small town in Arkansas called Figure Five. In 1908 he was in charge of the Adventist mission in Kagoshima. Young Irving Joshua Bush, who later took the name of Matrix, was the eldest of seven children, all but the first three born in Kagoshima. He grew up a devout believer in the biblical prophecies of his parents' faith, and, owing to a natural bent in mathematics, was particularly intrigued by the numerical aspects of those prophecies. At the age of seven he surprised his father by pointing out that there are 1 God, 2 testaments, 3 persons in the Trinity, 4 Gospels, 5 books of Moses, 6 days of creation, and 7 gifts of the Holy Spirit.

"What about 8?" his father had asked.

"It is the holiest number of all," the boy replied. "The other numbers with holes are 0, 6, and 9, and sometimes 4, but 8 has *two* holes, therefore it is the holiest."

At the age of eight young Bush devoted most of his spare time to investigating the numbers that occur in various biblical passages. For example, 1 Chronicles, chapter 20, verse 6, says that the giant of Gath had six toes on each foot and six fingers on each hand. It is no coincidence, the boy insisted, that 20, the number of the chapter, gives the normal allotment of toes and fingers to a man, and 6, the number of the verse, describes the abnormality of each hand and foot of the man of Gath. Moreover, said the boy, if we assign numbers to each letter in *Gath*, letting *a* equal 1, *b* equal 2, *c* equal 3, and so on, the numbers add to 36, the square of 6.

At the age of nine the budding numerologist applied a similar technique to his last name, *Bush*, obtaining the numbers 2, 21, 19, and 8. They coincided exactly with his birth date, the twenty-first day of the second month of the year 1908—an astonishing correlation, which he took to be a favorable omen about God's plans for him as a laborer in the Adventist cause.

In 1920, when Bush was thirteen, those plans were suddenly shattered. He found hidden in a dark corner of his father's study a copy of D. M. Canright's explosive book *Life of Mrs. E. G. White, Seventh-Day Adventist Prophet: Her False Claims Refuted* (Cincinnati: Standard Publishing, 1919). Shaken and disenchanted by the disclosures of this book, and finding himself in hopeless conflict with the fundamentalist views of his parents, he ran away from home, eventually making his way to Tokyo. He spoke, of course, fluent Japanese as well as English.

In addition to his early interest in numbers, young Bush had also made a hobby of magic and juggling. An elderly Japanese friend of his father, who had once been in show business, had taught him some elementary juggling and sleight of hand. In Tokyo he supported himself for several years by juggling and doing magic tricks on street corners. A famous Japanese magician named Tenkai saw him work and hired him as an assistant. Later, in his twenties, Bush traveled throughout Japan performing a mind-reading act under the stage name of Dr. Matrix. In 1938 he married his assistant, Miss Eisei Toshiyori, whose father was a Japanese foot juggler and trick bicycle rider. Their only child, a daughter, was born the following year.

Mrs. Matrix was killed in April 1942, during the bombing of Tokyo. After the war with Japan came to its abrupt end, Dr. Matrix took up residence in Paris, where, on the Left Bank, he quickly achieved a considerable reputation as an astrologer and numerological consultant. It is said, although I cannot vouch for it, that Charles de Gaulle once sought his advice on whether he should make André Malraux his information minister, and received a convincing affirmative answer based on a careful analysis of the birth dates and full names of the two men. It was while he was in Paris that Dr. Matrix became a personal friend of the world-renowned French mathematician Nicolas Bourbaki. Although Dr. Matrix had no formal schooling beyond the sixth grade at the mission school in Kagoshima, he had managed to teach himself a surprising amount of number theory. From the great

Bourbaki he acquired even deeper insights into this fundamental branch of mathematics.

I would have liked to include in this book a photograph of
Dr. Matrix and Iva, but alas, they would never allow me to take
a picture of either of them. As for Iva, I have not heard from her
since her father was killed. Perhaps she will read these words and
get in touch with me.

MARTIN GARDNER

Hendersonville, N.C.

1. New York

Numerology, the study of the mystical significance of numbers, has a long, complicated history that includes the ancient Hebrew cabalists, the Greek Pythagoreans, Philo of Alexandria, the Gnostics, many distinguished theologians, and those Hollywood numerologists who prospered in the 1920s and 1930s by devising names (with proper "vibrations") for would-be movie stars. I must confess that I have always found this history rather boring. Thus when a friend of mine suggested in late December 1959 that I get in touch with a New York numerologist who called himself Dr. Matrix, I could hardly have been less interested.

"But you'll find him very amusing," my friend insisted. "He claims to be a reincarnation of Pythagoras, and he really does seem to know something about mathematics. For example, he pointed out to me that 1960 had to be an unusual year because 1,960 can be expressed as the sum of two squares—14^2 and 42^2—and both 14 and 42 are multiples of the mystic number 7."

I made a quick check with pencil and paper. "By Plato, he's right!" I exclaimed. "He might be worth talking to at that."

I telephoned for an appointment, and several days later

a pretty secretary with dark, almond-shaped eyes ushered me into the doctor's inner sanctum. Ten huge numerals from 1 through 10, gleaming like gold, were hanging on the far wall behind a long desk. They were arranged in the triangular pattern made commonplace today by the arrangement of bowling pins, but which the ancient Pythagoreans viewed with awe as the "holy tetractys." A large dodecahedron on the desk had calendars for each month of the new year on each of its twelve sides. Soft organ music was coming from a hidden loudspeaker.

Dr. Matrix entered the room through a curtained side door; he was a tall, bony figure with a prominent nose and bright, penetrating green eyes. He motioned me into a chair. "I understand you write for *Scientific American*," he said with a crooked smile, "and that you're here to inquire about my methods rather than for a personal analysis."

"That's right," I said.

The doctor pushed a button on a side wall, and a panel in the woodwork slid back to reveal a small blackboard. On the blackboard were chalked the letters of the alphabet, in the form of a circle that joined Z to A (see Figure 1).

"Let me begin," he said, "by explaining why 1960 is likely to be a favorable year for your magazine." With the end of a pencil he began tapping the letters, starting with A and proceeding around the circle until he counted 19. The nineteenth letter was S. He continued around the circle, starting with the count of 1 on T, and counted up to 60. The count ended on A. S and A, he pointed out, are the initials of *Scientific American*.

"I'm not impressed," I said. "When there are thousands of different ways that coincidences like this can arise, it becomes extremely probable that with a little effort you can find at least one."

"I understand," said Dr. Matrix, "but don't be too sure

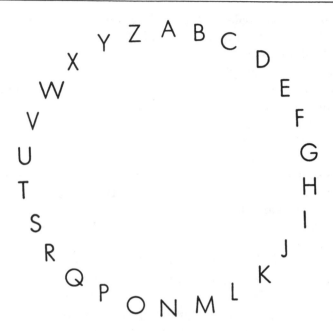

Figure 1. Dr. Matrix's alphabet circle

that's the whole story. Coincidences like this occur far
more often than can be justified by probability theory.
Numbers, you know, have a mysterious life of their own."
He waved his hand toward the gold numerals on the wall.
"Of course those are not numbers. They're only symbols for
numbers. Wasn't it the German mathematician Leopold
Kronecker who said: 'God created the integers; all the rest
is the work of man'?"

"I'm not sure I agree with that," I said, "but let's not
waste time on metaphysics."

"Quite right," he replied, seating himself behind the
desk. "Let me cite a few examples of numerological analysis that may interest your readers. You've heard, perhaps,

the theory that Shakespeare worked secretly on part of the King James translation of the Bible?"

I shook my head.

"To a numerologist, the theory's not in doubt. If you turn to Psalms 46 you'll find that its 46th word is SHAKE. Count back to the 46th word from the end of the same psalm [the word SELAH at the end is not part of the psalm—M.G.] and you reach the word SPEAR."

"Why 46?" I asked, smiling.

"Because," said Dr. Matrix, "when the King James Authorized Version was completed in 1610, Shakespeare was exactly 46 years old."

"Not bad," I said as I scribbled a few notes. "Any more?"

"Thousands," said Dr. Matrix. "Consider the case of Richard Wagner and the number 13. There are 13 letters in his name. He was born in 1813. Add the digits of this year and the sum is 13. He composed 13 great works of music. *Tannhäuser,* his greatest work, was completed on April 13, 1845, and first performed on March 13, 1861. He finished *Parsifal* on January 13, 1882. *Die Walküre* was first performed in 1870 on June 26, and 26 is twice 13. *Lohengrin* was composed in 1848, but Wagner did not hear it played until 1861, exactly 13 years later. He died on February 13, 1883. Note that the first and last digits of this year also form 13. These are only a few of the many important 13's in Wagner's life."

Dr. Matrix waited until I had finished writing; then he continued. "Important dates are never accidental. The atomic age began in 1942, when Enrico Fermi and his colleagues achieved the first nuclear chain reaction. You may have read in Laura Fermi's charming biography of her husband, *Atoms in the Family,* how Arthur Compton telephoned James Conant to report the news. Compton's first remark was: 'The Italian navigator has reached the New World.' Did it ever occur to you that if you switch the middle digits of 1942, it becomes 1492, the year that Columbus, an earlier Italian navigator, discovered the New World?"

"No," I answered, "I can't say it ever did."

"There's more. On that afternoon of December 2, 1942, in Fermi's laboratory under the University of Chicago's football stadium, exactly forty-two people were present when Fermi examined the dials and reported that the atomic reaction was self-sustaining." *

"Astonishing," I said, writing furiously.

"The life of Kaiser Wilhelm I is numerologically interesting," he went on. "In 1849 he crushed the socialist revolution in Germany. The sum of the digits in this date is 22. Add 22 to 1849 and you get 1871, the year Wilhelm was crowned emperor. Repeat this procedure with 1871 and you arrive at 1888, the year of his death. Repeat once

---

* The following letter, from the physicist Luis W. Alvarez, was published in *Scientific American* (Apr. 1960):

I enjoyed reading Martin Gardner's account of his visit with Dr. Matrix. When the doctor was discussing the first chain reaction, he was certainly on the right track, but because he did not work actively on the Manhattan District project, he missed some important verifications of his conclusions. He would have known, of course, that the only reason the pile was built during the war was to produce plutonium, the 94th element in the periodic system. What Dr. Matrix missed by not having Manhattan District clearance was the fact that the code designation for plutonium, all during the war, was "49." If the good doctor had had this fact available to him, he would also have pointed out that element 94 was discovered in California, the land of the 49'ers.

Since the real test of a new theory is its ability to predict new relationships which the author of the theory could not have foreseen, you have convinced me that numerology is here to stay.

more and you get 1913, the last year of peace before World War I destroyed his empire. Unusual date patterns are common in the lives of all famous men. Is it coincidence that Raphael, the great painter of sacred scenes, was born on April 6 and died on April 6, and that both dates fell on Good Friday? Is it a coincidence that Shakespeare was born on April 23 and died on April 23, and that twice 23 is 46, the number I mentioned before as the key to his work on the Bible?"

"And 23 is the number of the best-known Psalm," I added, "which presumably Shakespeare translated."

The doctor nodded and continued. "Exactly one hundred years ago three famous philosophers were born: John Dewey, Henri Bergson, and Samuel Alexander. For all three, evolution was the cornerstone of their philosophical visions. Why? Because 1859 was the year that Darwin's *Origin of Species* was published. Do you think it accidental that Houdini, the lover of mystery, died on October 31, the date of Halloween?"

"Could be," I murmured.

The doctor shook his head vigorously. "I suppose you'll think it coincidental that in the library's Dewey decimal system the classification for books on number theory is 512.81."

"Is there something unusual about that?"

"The number 512 is 2 to the ninth power and 81 is 9 to the second power.* But here's something even more remarkable. First, 11 plus 2 minus 1 is 12. Let me show you how this works out with letters." He moved to the blackboard and chalked on it the word ELEVEN. He added TWO to make ELEVENTWO, then he erased the letters of ONE,

---

* This had been pointed out by Harry Lindgren in the *Australian Mathematics Teacher* 8 (1952):8.

leaving ELEVTW. "Rearrange those six letters," he said, "and they spell TWELVE."

I dabbed at my forehead with my handkerchief. "Do you have any opinion about 666," I asked, "the so-called Number of the Beast [Revelation 13:18]? I recently came across a book called *Our Times and Their Meaning*, by a Seventh-Day Adventist named Carlyle B. Haynes. He identified the number with the Roman Catholic Church by adding up all the Roman numerals in one of the Latin titles of the pope: *Vicarius Filii Dei*. It comes to exactly 666."

V —	5
I —	1
C —	100
A —	
R —	
I —	1
U —	5
S —	
F —	
I —	1
L —	50
I —	1
I —	1
D —	500
E —	
I —	1
	666

(The letter *u* is taken as *v*, because that is how *u* was formerly written.)

"I could talk for hours about 666," the doctor said with a heavy sigh. "This particular application of the Beast's number is quite old.* Of course it's easy for a skillful numerologist to find 666 in any name. In fact, if you add the Latin numerals in the name ELLEN GOULD WHITE, the inspired prophetess who founded Seventh-Day Adventism—counting *w* as a 'double *u*' or two *v*'s—it also adds up to 666."

```
E —
L —    50
L —    50
E —
N —

G —
O —
U —     5
L —    50
D —   500

W—     10
H —
I —      1
T —
E —    ___
       666
```

"Tolstoy," Matrix continued, "in the first part of the third volume of *War and Peace*, chapter 19, has a clever method of extracting 666 from L'EMPEREUR NAPOLÉON.

---

* Finding 666 in the Roman numerals of *Vicarius Filii Dei* goes back at least to the seventeenth century. Early Adventists had other interpretations of 666, but this became the dominant one after Uriah Smith wrote in an 1866 (note the 66) magazine article that it was the "most plausible" explanation of 666 he had ever seen. In the first edition (1897) of his book, *Daniel and the Revelation*, he defends the interpretation at length and says that he first came upon it in an 1832 book on the Reformation. Although Mrs. White greatly admired Smith, she seems

When the prime minister of England was William Gladstone, a political enemy wrote GLADSTONE in Greek,
added up the Greek numerals in the name, and got 666.
HITLER adds up neatly to the number if we use a familiar
code in which a is 100, b, is 101, c is 102, and so on."

$$
\begin{array}{rl}
H- & 107 \\
I - & 108 \\
T - & 119 \\
L - & 111 \\
E - & 104 \\
R - & \underline{117} \\
& 666
\end{array}
$$

"I suppose you know," I said, "that 666 spots, like those
ten numerals on your wall here, can be arranged in triangular formation."

"Yes, 666 is a triangular number with a side of 36. It is
the sum of all the numbers on a standard magic square of
order 6. But here's something even more curious. If you
put down from right to left the first six Roman numerals, in
serial order, you get this."

He wrote DCLXVI (which is 666) on the blackboard.

"But what does it all mean?" I asked.

Dr. Matrix was silent for a moment. "The true meaning

---

never to have published her own view on the meaning of 666. That
Adventists still take Smith's interpretation seriously is evident from its
defense in the revised edition of Roy Allan Anderson's *Unfolding the
Revelation* (Mountain View, California, Pacific Press, 1974.)

I had occasion many years later to ask Dr. Matrix if he had invented
the application of the same technique to Mrs. White's name. He told
me no, he had found this in an anonymous tract published by the People's Christian Bulletin, New York City, perhaps in the 1930s. "The
Adventists," Dr. Matrix added, "missed a much better way of linking
the pope to 666. As my friend Raymond L. Holly once pointed out, the
first six words of Psalm 60, verse 6, are: 'God hath spoken in his holi-
ness.' "

For more on 666, see chapters 3, 4, 6, 12, 17, and 18.

is known only to a few initiates," he said unsmilingly. "I'm afraid I can't reveal it at this time."

"Would you be willing to comment on the coming presidential campaign?" I asked. "For instance, will Nixon or Rockefeller get the 1960 Republican nomination?"

"That's another question I prefer not to answer," he said, "but I would like to call your attention to some curious counterpoint involving the two men. NELSON begins and ends with *n*. ROCKEFELLER begins and ends with *r*. Nixon's name has the same pattern in reverse. RICHARD begins and almost ends with *r*. NIXON begins and ends with *n*. Do you know when and where Nixon was born?"

"No," I said.

"At Yorba Linda, California—in January 1913." Dr. Matrix turned back to the blackboard and wrote this date as 1–1913. He added the digits to get 15. On the circular alphabet he circled *Y*, *L*, and *C*, the initials of Nixon's birthplace, then he counted from each letter to the fifteenth letter from it clockwise to obtain *N.A.R.*, the initials of Nelson Aldrich Rockefeller! "Of course," he added, " of the two men, Rockefeller has the better chance to be elected."

"How is that?"

"His name has a double letter. You see, because of the number 2 in '20th century,' every president of this century must have a double letter in his name, like the *oo* in Roosevelt and the *rr* in Harry Truman."

I later verified that, with the exception of Dwight David Eisenhower, Matrix was correct. The presidents of the twentieth century, preceding Kennedy, were:

William McKinley
Theodore Roosevelt

Wi*ll*iam Taft
Woo*drow* Wilson
Wa*rr*en Harding
Calvin C*oo*lidge
Herbert H*oo*ver
Franklin R*oo*sevelt
Ha*rr*y Truman
Dwight Eisenhower

"Ike doesn't have a double letter," I said.

"Eisenhower is the one exception so far. We must remember, however, that he ran twice against Adlai Ewing Stevenson, who also lacks the double letter. Two things tipped the scales in Ike's favor: his double initials, *D.D.*, and the fact that *w* is simply an abbreviation for 'double *u.*' "

I glanced toward the blackboard. "Any other uses for that circular alphabet?"

"It has many uses," he replied. "Let me give you a recent example. The other day a young man from Brooklyn came to see me. He had renounced a vow of allegiance to a gang of hoodlums and he thought he ought to leave town to avoid punishment by gang members. Could I tell him by numerology, he wanted to know, where he should go? I convinced him he should go nowhere by taking the word ABJURER—it means one who renounces—and substituting for each letter the letter directly opposite it on the alphabet circle."

Dr. Matrix drew chalk lines on the blackboard from *A* to *N*, *B* to *O*, and so on. The new word was NOWHERE. "If you think that's a coincidence," he said, "just try it with even shorter words. The odds against starting with a seven-letter word and finding a second one by this technique are astronomical."

I glanced nervously at my wristwatch. "Before I leave, could you give me a numerological problem or two that I could ask my readers to solve?"

"I'll be delighted," he said. "Here's an easy one." On my notepaper he wrote the letters: *OTTFFSSENT.*

"On what basis are those letters ordered?" he asked. "It's a problem I give my beginning students of neo-Pythagoreanism. Please note that the number of letters is the same as the number of letters in the name Pythagoras." (See Answers, One, I.)

Beneath these letters he wrote:

$$
\begin{array}{r}
\text{F O R T Y} \\
\text{T E N} \\
\underline{\text{T E N}} \\
\text{S I X T Y}
\end{array}
$$

"Each letter in that addition problem stands for a different digit," he explained. "There's only one solution, but it takes a bit of brain work to find it." (See Answers, One, II.)

I pocketed my pencil and paper and stood up. Organ music continued to pour into the room. "Isn't that a Bach recording?" I asked.

"It is indeed," answered the doctor as he walked me to the door. "Bach was a deep student of our science. Have you read Leonard Bernstein's *Joy of Music?* It has an interesting paragraph about Bach's numerological investigations. He knew that the sum of the values of BACH—taking *a* as 1, *b* as 2, and so on—is 14, a multiple of the divine 7. He also knew that the sum of his entire name, using an old German alphabet, is 41, the reverse of 14, as well as the fourteenth prime number when you include 1 as a prime. The piece you're hearing is *Vor deinen Thron tret' ich allhier,* a hymn in which the musical form exploits this 14–41 motif. The first phrase has 14 notes, the entire melody has 41. Magnificent harmony, don't you think? If only our modern composers would learn a little numerology, they might come as close as this to the music of the spheres!"

I left the office in a slightly dazed condition; but not too dazed to notice again on my way out that the doctor's secretary had 1 upturned nose, 2 luminous eyes, and a most interesting overall figure.

2. Los Angeles

Dr. Matrix's prediction in 1959 that the next president of the United States would have a double letter in his name was, of course, dramatically confirmed in 1960 by the election of John Fitzgerald Kennedy. Indeed, Kennedy was the only serious contender for the Democratic nomination whose name contained a double letter, and his opponent in the election, Richard Milhous Nixon, lacked the double letter. As 1960 drew to a close, it occurred to me that Dr. Matrix might have some similar insights into the events of 1961, but when I tried to reach him by telephone, I discovered that early in the fall he had moved to Los Angeles. I finally established contact with him by mail and made arrangements to see him in December when I would have occasion to fly to L.A. for other reasons.

His office was on the fourteenth floor of a large apartment building not far from the University of California at Los Angeles. When I entered his spacious reception room I was pleased to see that Miss Iva Toshiyori, his attractive Eurasian secretary, was still with him. She gave me a big smile of recognition, and immediately ushered me into the numerologist's office.

"Please sit down," she said. "Dr. Matrix will be with you in a moment."

Soft strains of unfamiliar music, oriental in flavor, came from a concealed overhead source. Large gold numerals, in a square array, gleamed against a backdrop of black velvet on the wall behind a glass-topped desk. I was adding the rows to see if they formed a magic square when Dr. Matrix parted the curtains of a side door and entered the room.

We shook hands. His tall, lean figure towered over me, and his emerald eyes seemed to be observing me with sly amusement.

"On my last visit," I said, "those numerals formed the holy triangle of the Pythagoreans."

"Ah, yes," he said, turning his hawklike nose toward the back wall. "I like to change their arrangement every month. What you see here is an antimagic square. Each row, column, and main diagonal sums to a *different* total."

"Most interesting," I commented as I jotted the square on my note pad (see Figure 2).

"Not really," he replied. "There are several hundred ways to construct such a square. Much stronger antimagic

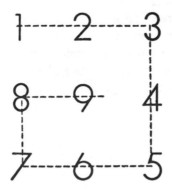

Figure 2. A rookwise-connected antimagic square

three-by-three squares are possible, with additional sets of three cells participating in the antimagic property, but I'm rather fond of this one because of the spiral way the digits are arranged."

"You might say," I remarked, "that the nine digits of the square are rookwise connected in consecutive order. If you put a chess rook on 1, you can move it along a path of standard rook moves [no diagonal moves permitted] so it visits all the digits in serial order from 1 to 9, without crossing over any it has already visited. Are there any other order-three antimagic squares with this same property?"

"Yes, but only one."

Dr. Matrix sketched it for me. Can the reader discover it? Rotations and reflections of patterns are not, of course, considered different. (See Answers, Two, I.)

"By the way," I said, "coming up in the elevator I noticed that the building has no thirteenth floor. Doesn't that make your floor—the fourteenth—really the thirteenth?"

Dr. Matrix's green eyes twinkled with amusement. "Of course. I hope you don't think me superstitious or a victim of triakaidekaphobia.* You may not be aware of it, but 13 is a ubiquitous number in the symbolism of the United States. Why, you Americans are surrounded with reminders of your 13 original states, from the number of stripes on your flag to the number of buttons on the flap of a sailor's dress blues. Do you have a dollar bill?"

---

* Triskaidekaphobia: An irrational fear of the number 13. For an amusing article on the superstition that if thirteen people dine at the same table a tragedy will befall one of them, see "Thirteen at Table," by Vincent Starrett, in *Gourmet* (Nov. 1966). It was one of Franklin D. Roosevelt's superstitions. His secretary Grace Tully, in *F.D.R., My Boss*, writes that FDR would often hastily summon her to attend a lunch or dinner because last-minute changes had brought the number at the table to thirteen. The article contains many other fascinating historical anecdotes and literary references.

I took a bill from my wallet; Dr. Matrix pointed to the green side, on which are reproduced the two faces of the Great Seal of the United States. "If you count the steps of the pyramid," he said, "you'll find there are exactly 13. The motto above the pyramid, *Annuit coeptis,* has 13 letters. The bald eagle on the right has a ribbon in its beak that bears the motto *E pluribus unum*—also 13 letters. Over the eagle's head are 13 stars. There are 13 stripes on the shield. The eagle's left talon (the right claw as you see it) holds 13 war arrows and its right talon holds an olive branch of peace with 13 leaves. At the base of the pyramid you'll see the date 1776 in Roman numerals. 7 and 6 add up to 13. Of course all this doesn't prevent me from trying to collect as many dollar bills as I can."

"Nor me," I agreed. "And I could have collected quite a few of them last January if I had followed your hints and bet on Kennedy's election."

"Yes," he said. "Kennedy's double *n* gave him enormous odds against all the candidates except Rockefeller. In addition, if we use the ancient Celtic numerological key, appropriate to Kennedy's Irish ancestry, the results are also propitious." Dr. Matrix pressed a button on his desk, and a panel on the wall slid back to reveal a large blackboard. He walked to the blackboard and chalked on it the digits from 0 through 9. Beneath these digits he printed the alphabet (see Figure 3).

"If we add the numerical values of J. F. KENNEDY," he explained, "we obtain 35, and of course the winner of the election is the thirty-fifth president."

"Have you tried this with R. M. NIXON?" I asked.

"Yes. It adds up to 30, the newspaper reporter's traditional symbol for the end of a story. Incidentally, I would like to emphasize that the double *n* in Kennedy's name marks an important break in the double-letter law. As C. C. Basore of Mountain Center, California, has pointed out,

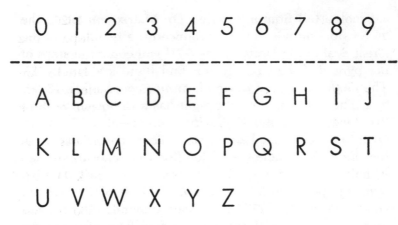

Figure 3.   An old Celtic key for analyzing the names of Kennedy
           and Nixon

there have been nine other presidents in this century with
double letters; in every case the letter is *l*, *o*, or *r*. If we
include the only exception, Dwight Eisenhower, on the
basis of the 'double *u*' in *w*, then the set of doubled letters
is *l*, *o*, *r*, and *u*. Note that these letters form an arithmetic
progression in the alphabet. Each is three letters ahead of
the one before it. Kennedy's double *n* is not part of that
progression. Perhaps this will provide his escape from another, rather ominous law."

"I know what you are referring to," I said. "Since 1840,
a year with digits that add to the unlucky 13, every president elected in a year ending in 0 has died in office."

(Harrison was elected in 1840. He died one month later.
There was, of course, no election in 1850. Lincoln, elected
in 1860, was assassinated. Garfield, elected in 1880, and
McKinley, elected in 1900, were also assassinated. Har-
ding—1920—and Roosevelt—1940—both died in office.
The only presidents elected before 1840 in years ending

with 0 were Jefferson in 1800 and Monroe in 1820. Neither man died in office, but both died on Independence Day.)

"Precisely," said Dr. Matrix. "It is closely related to another curious rule, which says that in every election year ending in 0 the elected president has either two or three syllables in his last name."

"I've heard about that also," I said. "There was an article about it last year in the *New York Times Magazine.*" The article was "Sibylline Syllables," by Jack Doherty (Feb. 22, 1959).

Dr. Matrix nodded. "From 1800 through 1860 the pattern of syllables was 3, 2, 3, 2. (In 1800, Jefferson; in 1820, Monroe; in 1840, Harrison; in 1860, Lincoln.) Then the pattern seemed to reverse itself and became 2, 3, 2, 3. (In 1880, Garfield; in 1900, McKinley; in 1920, Harding; in 1940, Roosevelt.) It looks as if 1960 marks a switch back to the original pattern of 3, 2, 3, 2. That would favor the three-syllabled Kennedy over two-syllabled Nixon and eliminate completely the four-syllabled Rockefeller."

"And Kennedy *did* win."

"Yes," said Dr. Martrix, "but the preceding pattern of three syllables is a bit forced. FDR always pronounced his name with two syllables—*Rose-velt.*"

After jotting all this down, I said, "Do you have any explanation of Rockefeller's famous slip of the tongue at the Republican convention, where he introduced Nixon as Richard *E.* Nixon?"

"Yes, indeed. Slips of the tongue are seldom accidental. Freud was right in attributing them to unconscious hopes and fears, but he underestimated the important role also played by numerical and verbal structure."

"Are you referring to the fact that Thomas Dewey's middle initial is *E.* ?"

"Partly that, but also much more." Dr. Matrix returned to the blackboard and chalked on it the initials *R.E.N.*

"It is true," he continued, "that Rockefeller's subconscious was linking Nixon with Thomas E. Dewey, the last Republican candidate to be defeated. But note what happens when we add Dewey's first and last initials." He printed a *T* in front of the three letters on the blackboard, and a *D* at the end, forming the word TREND.

"It was Rockefeller's subconscious," said Dr. Matrix, "expressing its hope that the national trend for Nixon would follow the same trend as it did for Dewey. But I think an even more important hidden attitude turns up if we take the first four letters of MILHOUS, Nixon's middle name, and see what happens when the *m* becomes an *e*."

Dr. Matrix printed *eilh* on the blackboard. "The election year was 1960, so 6 is our key number here. The sixth letter ahead of *e* in the alphabet is *k*, the sixth letter ahead of *i* is *o*, the sixth letter ahead of *l* is *r*, and the sixth letter ahead of *h* is *n*." As he spoke he chalked above *eilh* the letters *korn*. "This has a double meaning," he went on. "First, it expressed Rockefeller's private opinion of the speech he had just made; second, it expressed his secret conviction that Kennedy would 'K.O.' *R.N.*"

"And you seriously believe all this is more than coincidence?" I asked.

"Yes, I do," answered Dr. Matrix, unsmiling. "It is naive to suppose that there is such a thing as a randomly arranged group of symbols. Random means without order or pattern. The term obviously is self-contradictory. You can no more find a patternless arrangement of digits or letters than you can find a cloud without a shape or a culture without folkways. In my opinion every pattern of symbols conceals a secret meaning, though it may require great skill to discover it. Death dates, in particular, are often correlated with earlier patterns. 'In today,' as Schiller so aptly phrased it, 'already walks tomorrow.' I could give you thousands of examples. Dickens, in the last paragraph of his last completed novel, *Our Mutual Friend*, tells how he narrowly escaped being killed in a railway accident on June 9. Five years later, on June 9, Dickens died. Have you ever noticed the tendency of great political events to occur on strongly patterned dates? The bugles sounded cease-fire at the close of the First World War at the eleventh hour of the eleventh day of the eleventh month, 1918. The invasion of France by the Allies in the Second World War began at the sixth hour of the sixth day of the sixth month, 1944. Roosevelt, Churchill, and Stalin opened their famous Yalta meeting on February 3, 1945, a date that has four digits in serial order when it is written 2–3–45. West Germany became a sovereign state on 5–5–55. Stalin died on March 5, 1953, a date with the interesting mirror reflection of 3–5–53."

"Are there any patterned dates in 1961?"

"How could it be otherwise? The two most striking date patterns are 1–6–61 and 6–1–61, or January 6 and June 1. Events of worldwide import have a high probability of taking place on both occasions. February 8 is another date that bears close watching; if this is written out in full it is 2–8–1961, and 281,961 is the square of 531, or May 31. I anticipate important, closely linked events on both those days. The year itself will surely be a topsy-turvy one. As you no doubt already know, it's the first year since 1881 that is the same upside down. There won't be another year like that until 6009."

Dr. Matrix waited until I finished scribbling, then he continued. "It is in the physical world, however, that we find the most remarkable numerical patterns. A numerologist knows it is not a coincidence that the sun's disk, viewed from the earth, is almost identical in size to the moon's disk,* or that the sun's period of rotation is almost exactly the same as the moon's period of revolution around the earth. Has it ever occurred to you how strange it is that there are 365 days in the year?"

I shook my head.

"It's a truly amazing instance of divine harmony—the harmony that Kepler so clearly perceived, but which later astronomers have regrettably ignored. 365 is not only the sum of 10 squared plus 11 squared plus 12 squared; it is also the sum of 13 squared and 14 squared. We can write it this way."

He printed on the blackboard:

$$10^2 + 11^2 + 12^2 = 13^2 + 14^2 = 365$$

"That solution is the second example of an infinite set. Everyone knows the first example—3 squared plus 4 squared equals 5 squared—in which the integers are in consecutive order, with two terms on the left and one on the right. The example on the blackboard has three terms on the left, two on the right. The third example, with four terms on the left, three on the right, is—" and he wrote on the blackboard:

$$21^2 + 22^2 + 23^2 + 24^2 = 25^2 + 26^2 + 27^2$$

---

* This is dramatically evident during a total eclipse of the sun when the moon's disk precisely occludes the sun's disk. Put another way, the tip of the moon's shadow just brushes the earth's surface. The improbability of this coincidence is a cornerstone in a proof of the existence of God, as outlined in a ten-page pamphlet by Norman Bloom, published in 1970 in Guttenberg, New Jersey. The pamphlet is entitled *The New World: The First Proof in History That the Earth, Moon and Sun Are Controlled by a Thinking, Acting Mind and Hand That Has the Power of Life and Death over Every Living Thing on Earth.* Mr. Bloom reports that he has defended his argument at such centers of higher learning as Harvard, MIT, and Barry Farber's WOR radio show. He offers $1,000 to anyone who can find a flaw in his proof.

"You might ask your readers to see if they can find the fourth example, with five terms on the left, four on the right, and perhaps give a simple formula for finding all the higher examples." (See Answers, Two, II.)

Dr. Matrix was silent until I finished writing this down, then he asked, "Are you familiar with Arthur Stanley Eddington's work on the so-called fine-structure constant?" *

"Vaguely. The number is 137, isn't it? As I recall, Eddington had a clever way of deducing it, apart from experimental observation. Didn't he first arrive at 136?"

Dr. Matrix nodded. "He gave a complicated mathematical explanation of why he revised it to 137, but the truth is that Stanley was one of my most distinguished pupils. We worked it out one day over a bottle of Greek wine. We took Eddington's birth year, 1882, multiplied the digits to obtain 128, then added 9, the number of letters in Eddington's name."

"I can believe it," I said, chuckling. "Tell me, do you have any interesting numerological puzzles that my readers might enjoy?"

Dr. Matrix scratched his large nose. "Yes, I had an unusual problem called to my attention recently by my friend Dennis Sciama, a cosmologist at Cornell University. Suppose we wish to form a chain of symbols, using only the digits 1 and 2. How long a chain can we write without repeating a pattern, side by side? For example, we can't write 11 or 22, because each repeats the pattern of a single

---

* For an amusing account of numerological speculations about the still-unexplained fine-structure constant, including Eddington's "brave attempt" to derive it from a square matrix with a side of sixteen cells, see George Gamow, *Biography of Physics* (New York: Harper Torchbook, 1964), pp. 324–329. Dr. Matrix called my attention to the curious palindromic symmetry of the six digits that keep repeating between double zeros in .007299270072992700 . . . , the decimal for $1/137$. For biblical references to 137, see chapter 18.

digit, so we have to write, say, 12. The next digit has to be
1, but now we're stuck. We can't add another 1, and we
can't add 2, because it would repeat the two-digit pattern
of 12."

"In other words," I said, "the longest chain that can be
formed with two symbols, without having two adjacent
patterns that are duplicates, is a chain of three symbols."

"Correct. Now the problem is this. What is the longest
chain that can be written with *three* symbols? For example, we can't write 132132 because it repeats the pattern 132. But we can write 1323132, because now the two
patterns of 132 are separated. Is there a limit to the size of
a chain that can be constructed on this basis?" (See Answers, Two, III.)

On my way out I stopped to chat with Miss Toshiyori. "I
expect to be in L.A. for a few more days," I said. "Is
there any chance you could have dinner with me tomorrow night?"

She stopped typing and smiled. "Why don't you phone
me at home this evening? My number is . . ."

I whipped out my pad and pencil, wrote down the two
letters she gave me, then waited.

"The number has five digits," she said. "If you add a 4
to the front of it you make a number that is exactly four
times the number you make if you put the 4 at the end instead of the front."

I looked blank. "You mean I have to figure this out before I can call you?"

She nodded and started typing again. I figured it out all
right. Perhaps it will be less difficult for the reader. (See
Answers, Two, IV.)

3. Sing Sing

When I visited Dr. Matrix in December 1960 he had called attention to such "pattern dates" in 1961 as 6–1–61 (June 1, 1961) and had predicted that events of worldwide import would occur on those days. To my great disappointment, absolutely nothing of earthshaking consequence seemed to take place on any of the days the doctor had singled out. I wrote to him in June, after the last patterned day had slipped uneventfully by, to ask him how he could explain these apparent failures. The envelope was returned to me marked "Moved—left no forwarding address." All my efforts to locate him were unsuccessful.

Then, late in 1962, on my birthday (October 21), I received a postcard from him. "May the Fates be kind," it read, "to you on your 16/33 birthday." This puzzled me for some time, but when I divided 16 by 33, I found the quotient was a decimal that endlessly repeated my age. (Does the reader know how to arrive at such fractions? What, for example, is the smallest integral fraction that in decimal form endlessly repeats 27? See Answers, Three, I.)

The card bore no return address, but I was surprised to see that it had been postmarked in Ossining, a town on the east bank of the Hudson River about ten miles north of Dobbs Ferry, the town where I then lived. I leafed eagerly

through the Westchester County telephone book. No Dr.
Matrix. I tried Toshiyori. Yes, there she was! A moment
later I was speaking with her on the telephone.

The story she told was a sad one. After the failure of Dr.
Matrix's predictions, his clientele in Los Angeles had
slowly dwindled and he found himself sliding deeper and
deeper into debt. In desperation he did a foolish thing. He
tried to make some twenty-dollar bills.

His method was bizarre and surprising. With a paper
cutter he sliced each of fourteen bills into two parts, cutting them neatly along the broken vertical lines on each of
the schematic bills shown on the left side of Figure 4. The
right-hand portion of each bill was then butted against the
left-hand portion of the bill immediately following it. In the
process each bill lost more than it gained. The result: fifteen
bills! Each of them was only fourteen-fifteenths as long as a
legitimate bill, and all but the two end bills were glued
together along their cut edges, as indicated by the solid
vertical lines in the bills on the right side of the picture. The
loss of length was scarcely noticeable, and the carefully
glued butted edges formed only a barely perceptible hair-
line.

Unfortunately—or rather, fortunately—the United States
government places duplicate serial numbers at opposite
corners of every bill, and most of the numerologist's new
bills therefore bore pairs of serial numbers that did not
match. True, Dr. Matrix's method of making new bills was
not exactly counterfeiting—he merely "rearranged" the
parts of genuine bills.* Nevertheless, the Treasury Depart-

---

* This curious method of counterfeiting is an old one. Sam Loyd, who
based his famous "Get off the earth" paradox on the same principle,
wrote in his puzzle column in the *Brooklyn Daily Eagle* (Jan. 3, 1897,
p. 22): "The counterfeiter's trick of gradually converting twelve bills
into thirteen is closely allied to getting off the earth . . ." It has long

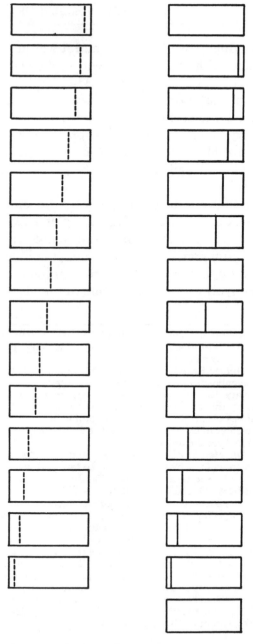

Figure 4. Fourteen bills (left) can be cut and rearranged to make fifteen (right). Each bill loses one-fifteenth of its length

ment took a dim view of his work and it was not long until he found himself firmly confined within the matrix of cells at Sing Sing. Sentence: five years. Miss Toshiyori took an apartment in nearby Ossining. She was allowed to visit Dr. Matrix twice a week, and with his assistance she was managing to carry on his numerological practice by mail.

"Yes," she said on the phone, "I'm sure I can arrange for you to see him. I'll call you back in a few days and let you know when."

It was a sunny winter afternoon—not a cloud was in sight—when I drove to Ossining and wound my way down the sloping side streets to the bank of the Tappan Zee. Behind the grim gray walls of Sing Sing the blue waters of the Hudson rippled pollutedly in the sunlight.

Miss Toshiyori was waiting in one of the visitor's rooms,

---

been a federal offense (although Dr. Matrix, unaccountably, was confined to a state prison) and those who are foolish enough to try it are easily and quickly caught.

The *Des Moines Register* (Aug. 20, 1963) reported that an eighteen-year-old Davenport youth, who had read my interview with Dr. Matrix in the January 1963 issue of *Scientific American*, had stupidly attempted to try this method of making money to get cash for entering college. After one of his phony twenty-dollar bills was spotted by a drugstore cashier in Davenport, police had little trouble tracing it to its source and making the arrest.

The *Boston Herald* (Feb. 28, 1963) and the *Harvard Crimson* (March 2) warned local merchants that fake tens and twenties made by this method were appearing in the Cambridge area. The *Chicago Daily News* (Nov. 29, 1968) reported that Reuben Silver of London had been given an eight-year prison term for cutting up five-pound notes to make twelve bills from eleven. "The court said the system was so good that details can't be made public." In the U.S., the method violates chapter 25, title 18 of U.S. Code section 484.

For a discussion of the many mind-bending geometrical paradoxes that operate on the same principle as this counterfeiting method, see chapters 7 and 8 of my *Mathematics, Magic and Mystery* (New York: Dover, 1956), and my *Scientific American* column on advertising premiums (Nov. 1971). The best and funniest printed puzzle of this type, the *Vanishing Leprechaun*, is obtainable from W. A. Elliott Co., 212 Adelaide W., Toronto, Canada M5H 1W7.

as enigmatic and beautiful as ever. She seemed unembar-
rassed by the context of our meeting. Through a grille beside her chair I recognized the hawklike nose and glittering green eyes of her employer.

Dr. Matrix was in a tired, unsmiling mood, but his voice
was cordial. "I'm afraid 1963 is a hopelessly dull number,"
he said. "Its reversal, 3691, is a prime, but so is the reversal for 1964. Too bad we didn't get together last January. I
could have told you that 987 plus 654 plus 321 is exactly
1962."

"What an astounding coincidence!" I exclaimed. "The
nine digits are in descending order!" *

Dr. Matrix shook his head. "Such things are never coincidental. They are part of that mysterious order in which
both mathematics and history lie embedded. Last year was
the year of the big countdown. Never before in history
have so many giant rockets been fired to the accompaniment of so many backward-recited integers."

He waited until I had jotted down his remarks in my
notebook. "Have you ever noticed that 12 is equal to 3
times 4, and 56 is equal to 7 times 8?"

I thought this over for a moment and gave a start when I
realized that Dr. Matrix's remark contained the first eight
digits in ascending order.†

"My number as a prisoner is rather interesting," he went
on. "It is 54748, a number of five digits. If you add

---

* This had been discovered by Charles Trigg and published in *Recreational Mathematics Magazine* (Apr. 1962, p. 33).

† This curiosity was contributed by Everett W. Comstock to *Recreational
Mathematics Magazine* (Apr. 1962, p. 36). There is no other set of four
consecutive numbers such that the first two, in ascending order, form
an integer equal to the product of the second two. If only each pair of
numbers need be consecutive, many equations can be written, such as
6,162 = 78 × 79. For equalities of this type, see J. A. Lindon's note in
the same magazine (Oct. 1962, p. 35).

the fifth powers of each digit [$5^5 + 4^5 + 7^5 + 4^5 + 8^5$], the sum is 54,748. I consider this a favorable omen."

"Are there many numbers like that?"

"Very few. The smallest is 153. It has three digits, so in this case we raise each digit to its third power. Sum the powers and you have 153 again. It is no accident that we are told in the last chapter of the Gospel of St. John, verse 11, that the net Simon Peter drew from the Sea of Tiberias contained 153 fish. The number has many mystical properties." (I later learned that only three other numbers are equal to the sum of the cubes of their digits. Each is a three-digit number. Can the reader discover them? See Answers, Three, II.)

"I seem to recall," I said, "that St. Augustine somewhere gives an elaborate numerological analysis of why those fish numbered 153."

"Yes, St. Augustine starts with 10, the number of the commandments and a symbol of the old Mosaic dispensation. To it he adds 7, the number of the gifts of the spirit and a symbol of the new dispensation. The resulting number, 17, signifies the union of old and new. He then sums the integers from 1 through 17 to obtain 153. Rather primitive numerology, in my opinion, but of course St. Augustine did not have the benefit of today's numerological techniques." *

---

* A remarkable property of 153 was discovered by Phil Kohn, Yokneam, Israel, and reported by Thomas H. O'Beirne in his column "Puzzles and Paradoxes," *New Scientist* (Dec. 21, 1961). Start with any integer that is a multiple of 3. Sum the cubes of its digits to obtain a second number. Keep repeating this procedure. In a finite number of steps you will arrive at the dead end of 153. O'Beirne gives a proof. Perhaps it is also of interest to observe that $153 = 1! + 2! + 3! + 4! + 5!$

If you start with any natural number not a multiple of 3, and keep summing the cubes of the digits, you eventually reach one of these dead ends: 1, 370, 371, 407, or one of the following repeating cycles: 55–250–133, 160–217–352, 136–244, 919–1,459. See problem E1810 in

"What about the number of your cell?" I asked.

Dr. Matrix smiled. "It has two digits. Put a decimal point between them and the new number is the average of the two digits. I'm sure your readers will find it a simple matter to determine the cell number. You might also ask them to compute the dimensions of my cell floor. It's a rectangle, not a square. I measured it one day and found that the length and width were each an integral number of yards. I realized immediately that the perimeter in yards exactly equals the area in square yards."

"Excellent," I said. "I'll give both problems." (See Answers, Three, III and IV.)

"We live in perilous times," Dr. Matrix continued. "Times that call for great wisdom on the part of our leaders. By the way, don't you think it odd that two of the leading philosophers of England should bear the name Wisdom: John Oulton Wisdom and Arthur John Terence Dibben Wisdom?"

"I understand they're cousins," I said.

"I'm glad to hear that," said Dr. Matrix. "It makes it less frightening. Have you read much of Franz Kafka? We are all trapped, you know, inside an insane labyrinth, like the protagonist 'K.' in Kafka's *The Castle*. That letter k, in my opinion, is one of the great prophetic symbols of modern literature. It is the eleventh letter of the alphabet, a symbol of the eleventh hour of the world. Who are the leaders of those two giant nations that now glower at each other around the sides of the planet?"

I must confess that little fingers tapped up and down my backbone when I realized that both KENNEDY and KHRUSHCHEV begin with k.

American Mathematical Monthly (Jan. 1967, pp. 87–88), and references there cited; and a second solution (March 1968, p. 294). For more on the 153 fishes, see chapter 18 below.

"We live today," Dr. Matrix continued wearily, "in the shadow of the H-bomb. The age of fission has become the age of fusion. FUSION is a word heavily charged with meaning for the numerologist."

"How is that?"

"You recall my alphabet circle?" He asked for my pencil and a sheet of paper. On the paper he jotted down the letters of the alphabet in circular form (see Figure 1), the Z joining the A like a snake with its tail in its mouth, the ancient symbol of eternal recurrence. "FUSION has six letters," he said. "We count clockwise around the circle, six letters from F. The count ends on L, the first letter of a new word. We do the same with U to obtain A, the second letter of the new word."

He continued in the same way with the remaining letters. When he finished, FUSION had been transformed to LAYOUT. "You see," he said, smiling crookedly, "the word FUSION leads inexorably, by way of 6, to an ominous hint of its power to flatten civilization."

"The world was created in six days," I said. "Didn't St. Augustine argue that God chose that number because it was the first of the so-called perfect numbers?" (A perfect number is one that is the sum of all its divisors: e.g., $6 = 1 + 2 + 3$.)

Dr. Matrix nodded. "But 6 is also the building block of 666, the mark of the Beast. The world today can be destroyed in six minutes."

"You seem as much interested in letters as in integers," I said, trying to change the subject. "My readers like occasional word puzzles. Any clever word problem you can give me?"

Dr. Matrix put the tips of his fingers together, leaned back and closed his eyes. "Hm. Let's see. Oh, yes, here's an amusing anagram that will no doubt annoy many of your readers. Simply rearrange the letters of CHESTY to form another English word. It isn't easy."

"A familiar word?"

"Yes indeed. In fact, a word most appropriate for the beginning of a new year." (See Answers, Three, V.)

When I asked Dr. Matrix about his life in prison, he made a wry face. "The name Sing Sing, as you may know, derives from the old tribe of Sint Sink Indians. We're all sin sick here. But I'm treated decently, I must admit. I work part time in the prison library. The roof of the recreation room leaks and is about to collapse, but we've managed to shore up one of its main beams. The food could be worse. Of course I never eat beans."

His remark puzzled me until I remembered the old Pythagorean prohibition about beans. "You consider yourself a member of the Pythagorean brotherhood?"

"My dear Gardner," he said, sitting more erect in his chair, "I *am* Pythagoras. I'm his eleventh reincarnation."

I exchanged amused glances with Miss Toshiyori, who had been listening silently to our conversation. "I suppose," I said, "you'll be telling me next that, like Pythagoras, you have a golden thigh."

Dr. Matrix said nothing. But he tapped the lower part of his thigh with the end of my mechanical pencil. It made a sharp metallic sound!

As Miss Toshiyori and I were leaving the prison's main gate, I asked: "Does Dr. Matrix really believe he's a reincarnation of Pythagoras?"

"Heavens, no," she said with a laugh. "But the old faker likes to keep up the pretense. What is it that show people say? He's always 'on.' It's part of his act."

"You're very loyal to him."

"He's my father."

My eyebrows hopped upward. "And your mother?"

She slipped her arm through mine. "Japanese. They met in Tokyo." Suddenly her dark eyes brightened. "I know a quaint cocktail lounge a few miles from here, in Tarrytown. If you'll buy me a martini, I'll tell you all about it."

At some future time, perhaps, I'll give her story. For now, I mention only one bewildering thing she said; it provides a fitting little problem with which to close this chapter.

When I asked her how old she was, she smiled and said cryptically: "The day before yesterday I was twenty-two, but next year I'll be twenty-five."

Can the reader deduce Miss Toshiyori's birthday, as well as the date on which our conversation took place? (See Answers, Three, VI.)

4. Lincoln and Kennedy

Had Dr. Matrix served his full term in Sing Sing he would not have been released until 1966. However, because of his model behavior, and as a reward for some hush-hush cryptographic services rendered to the State Department, he was given a parole in the fall of 1963. Washington refused to divulge details about his cryptographic work, though I suspect the doctor's vast knowledge of modern oriental languages was involved.

On November 20 I received a letter from Iva informing me of her father's parole a few months earlier. The two were living in Chicago's near North Side, at the Berkshire Hotel on East Ohio Street. The old mountebank, unable to reestablish a profitable clientele as a private numerological consultant, was seriously planning to enter show business. During his youth in Tokyo, Iva informed me, he had been a professional magician and stage mentalist. He was now preparing an unusual nightclub act in which numerology would be combined with mind reading and fortune-telling. She promised to let me know later how it all worked out.

On November 22 the president of the United States was assassinated.

A week later I received from Dr. Matrix a remarkable

letter. I did not release this letter to the press, but I made the mistake of sending a few copies to friends. By the spring of 1964, portions of the letter, in mimeographed or photocopied form, were making the rounds of offices in Manhattan, Washington, and elsewhere. They were published in *Newsweek* (Aug. 10) and *Time* (Aug. 21). As the statements were copied they were frequently garbled or misquoted. Here for the first time is the complete text of Dr. Matrix's original letter, exactly as I received it:

DEAR MARTIN GARDNER:

The two most dramatic and tragic deaths in American political history were the deaths of Abraham Lincoln and John Fitzgerald Kennedy. There are so many astonishing numerological parallels involving these two events of infamy that I am impelled to record them for you. Please use the following analysis as you wish, but with discretion.

1. Lincoln was elected president in 1860. Exactly one hundred years later, in 1960, Kennedy was elected president.

2. Both men were deeply involved in civil rights for Negroes.

3. Both men were assassinated on a Friday, in the presence of their wives.

4. Each wife had lost a son while living at the White House.

5. Both men were killed by a bullet that entered the head from behind.

6. Lincoln was killed in Ford's Theater. Kennedy met his death while riding in a Lincoln convertible made by Ford Motor Company.

7. Both men were succeeded by vice-presidents named Johnson who were southern Democrats and former senators.

8. Andrew Johnson was born in 1808. Lyndon Johnson was born in 1908, exactly one hundred years later.

9. The first name of Lincoln's private secretary was John, the last name of Kennedy's private secretary was Lincoln. [In pirated copies of parts of Dr. Matrix's letter this was usually altered to the incorrect assertion that Lincoln's private secretary was named Kennedy. His name was John Nicolay.—M.G.]

10. John Wilkes Booth was born in 1839. Lee Harvey Oswald was born in 1939, one hundred years later. [I have since learned that there is a controversy over Booth's birth date. Recent scholarship assigns the date 1838, and this apparently is correct. The 1839 date is given, however, by such standard references as *Chambers's Biographical Dictionary* (1962 ed.) and Funk & Wagnalls' *New Standard Dictionary of the English Language* (1945 ed.).—M.G.]

11. Both assassins were Southerners who held extremist views.

12. Both assassins were murdered before they could be brought to trial.

13. Booth shot Lincoln in a theater and fled to a barn. ["Barn" was changed to "warehouse" on most copies of Dr. Matrix's letter.—M.G.] Oswald shot Kennedy from a warehouse and fled to a theater.

14. LINCOLN and KENNEDY each has seven letters.

15. ANDREW JOHNSON and LYNDON JOHNSON each has thirteen letters.

16. JOHN WILKES BOOTH and LEE HARVEY OSWALD each has fifteen letters.

Both the Federal Bureau of Investigation and the Secret Service, had they been skilled in the prophetic aspects of numerology, would have been more alert on the fatal day.

The digits of 11/22 (November 22) add to 6, and FRIDAY has six letters. Take the letters *FBI*, shift each forward six letters in the alphabet, and you get *LHO*, the initials of Lee Harvey Oswald. He was, of course, well known to the FBI. Moreover, OSWALD has six letters. Oswald shot from the sixth floor of the building where he worked. Note also that the triple shift of *FBI* to *LHO* is expressed by the number 666, the infamous number of the Beast.

The Secret Service, an arm of the Treasury Department, likewise should have been more alert. Two weeks before the assassination the Treasury Department released a new series of dollar bills, a sample of which I enclose. [Dr. Matrix had pinned a dollar bill to his letter. See Figure 5.]

Observe that this series is designated by the letter *K* on the left. In 1913, half a century earlier, when the Federal Reserve Districts were designated, Dallas was assigned the letter K, the eleventh letter of the alphabet. For this reason "Dallas, Texas," where Kennedy was murdered, appears beneath the *K*. DALLAS, TEXAS has eleven letters. JOHN KENNEDY has eleven letters.

The serial number on this bill, as on all K bills, begins with *K* and ends with *A*—"*Kennedy A*ssassination." Beneath the serial number on the right is "Washington, D.C.," the origin of the President's fatal trip.

Below the serial number on the right, and above and below the serial number on the left, are pairs of 11's. Eleven, of course, is the month of November. The two 11's add to 22, the day of the assassination. To the right of Washington's picture is "Series 1963A," the year of the assassination.

"Thou [God] hast ordered all things in measure and number . . ."—The Wisdom of Solomon, *Old Testament Apocrypha*, chapter 11, verse 20.

Your humble servant,
[signed] IRVING J. MATRIX

Note added to page proofs: A letter from Terry W. Harmon has called my attention to another remarkable coincidence involving Lincoln and Kennedy. The first public suggestion that Lincoln be the Republican candidate

Figure 5.   The "Kennedy assassination bill"

for president is believed to be a letter written November 6, 1858, and published in the Cincinnati *Gazette*. The writer, Israel Green (a druggist in Findlay, Ohio), proposed a ticket of Lincoln for president, and John Kennedy for vice president.

The proposed Kennedy was John Pendleton Kennedy of Maryland, a prominent author and politician who had been Millard Fillmore's Secretary of the Navy. Green's letter can be found in *The Magazine of American History*, Vol. 29, 1893, pages 282–283.—M. G.

5. Chicago

Early in December 1963 I received a postcard from Iva informing me that her father had been booked into the Purple Hat Club and would open his numerological mind-reading act there on Saturday evening, December 14. I wired back immediately that I was unable to attend the opening, but would be in Chicago on the following Wednesday, December 18.

The Purple Hat is a popular cabaret in the Rush Street section of Chicago's near North Side. One can see its floor show from the bar. I arrived shortly before Wednesday night's first show, found an empty stool, ordered Scotch and water, then rotated 180 degrees to survey the scene.

Behind a small stage, at one end of the dance floor, was a black backdrop on which large numerals formed the order-3 square shown in Figure 6. Dr. Matrix had clearly intended to indicate the coming year by the way he had placed the digits of 1964. I was trying to determine what curious properties the matrix possessed when the lights began to dim and the Purple Hatters, a small group of black musicians wearing lavender top hats, began to play soft strains of oriental music.

A spotlight followed Dr. Matrix from a side entrance to the center of the stage. He was in full evening dress, tall and unsmiling, his green eyes glittering ominously above

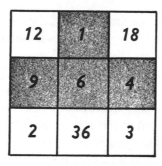

Figure 6.   Dr. Matrix's magic multiplication square

his prominent, convex nose. An enormous blood-red jewel glowed at the front of his snow-white turban. He bowed with a maximum of dignity, then turned to introduce his assistant, Miss Toshiyori. Iva bowed with a minimum of costume. A blue spotlight centered on her navel, she began to rotate her pelvis in a slow belly dance, the gyrations punctuated by the pounding rhythm of the Purple Hatters.

"Holy Muhammad!" whispered the man sitting on my left (a New York City booking agent, I learned later), "where did the old buzzard find *her?*"

After her dance was completed and the applause had subsided, Iva left the platform and slithered gracefully around the tables, gathering green cards on which patrons had been asked to write questions and sign their full name and date of birth. She tossed the cards into a big glass bowl and poured a purple liquid over them. Dr. Matrix muttered something to her in Japanese, then snapped his fingers. The contents of the bowl burst instantly into violet flames. Speaking in low, guttural tones, the dancing flames casting an eerie purple glow over his features from below, he began his "readings."

It was a brilliant performance. Shrewd guesses and predictions were cleverly interwoven with anagrams and other word plays on the patrons' names or initials, with curious numerological speculations involving birth dates, and with a masterful display of what is known in the trade as "cold reading." (A cold reading is a reading given without prior information about a subject. "Mrs. C.G.," Dr. Matrix would say, "the vibrations of your handwriting suggest that you recently had a phone call that was most disturbing." Mrs. C.G. would scream with astonished confirmation.) He received a good hand when he finished.

The lights went on. I saw Iva—she had slipped into a dress—approaching the bar. "Hi!" she said. "I thought I recognized you during the act. Come along. We have two hours until the next show."

I slapped some change on the bar. "Old friend," I explained to the bug-eyed man on my left.

The three of us took a taxi to Dr. Matrix's rooms, now at the more expensive Allerton Hotel. He was surprisingly cordial. "Yes," he said, "I chose that order-3 square because it contained 1964 so symmetrically. It is the simplest magic multiplication square. The product of the three numbers in any row—horizontal, vertical, or diagonal—is 216. It is the lowest possible product for such a square, assuming, of course, that each cell contains a different positive integer."

He paused until I had finished jotting down this information. "There is a pretty puzzle connected with this square. You might ask your readers to see if they can rearrange the same nine numbers to make a magic division square."

When I looked puzzled, he explained: The two end numbers of any line of three are multiplied, then the product is divided by the middle number. The final result must always be the same. (See Answers, Five, I.)

"Excellent," I said. "Any other curiosities connected with 1964?"

Dr. Matrix nodded gravely. "From a military point of view, the year is potentially explosive. The Boer War ended in 1902. If we add each of its four digits to 1902, we obtain 1914, the year World War I began. That war ended in 1919. Add 1, 9, 1, and 9 to 1919 and the result is 1939, the start of World War II."

"I dig it," I said. "World War II ended in 1945. Let's see . . ." I added 1, 9, 4, and 5 to 1945. The result: 1964!

"It would be foolhardy," said Dr. Matrix, "to ignore this obvious pattern. Of course, the numbers merely impel, they never compel."

"It occurred to me," I said, "that because the new year ends with 4 it might be an appropriate time to introduce my readers to the old pastime of the four 4's. Do you know the game?"

Dr. Matrix sighed painfully. "I know it well."

Let me first explain the recreation. One seeks to form as many whole numbers as possible, starting with 1, by using only the digit 4 four times—no more, no less—together with simple mathematical symbols. Naturally one must establish what is meant by a "simple" symbol. This traditionally includes the arithmetical signs for addition, subtraction, multiplication, and division, together with the square-root sign (repeated as many finite times as desired), parentheses, decimal points, and the factorial sign. (Factorial n is written $n!$ It means $1 \times 2 \times 3 \ . \ . \ . \ \times n$.) Two 4's can be represented as 44, three as 444, and so on. A decimal point may also be placed above .4, in which case it indicates the repeating decimal .4444 . . . , or 4/9.

The numbers 1 through 10 are easily expressed, in many different ways, by using no more than the symbols for multiplication, division, addition, and subtraction (see Figure 7). By adding the square-root sign, numbers 11 through 20 (except for 19) are readily obtained. By allowing the factorial sign and the dot used as both a decimal point and a repeating decimal sign, one can go on to 112. There seems to be no way to express 113 within these restrictions unless one employs highly bizarre combinations of the above symbols, such as the combined square root, decimal, and repeated decimal signs in the denominator of the first term in the following equation:

$$\frac{4!}{.\overline{\sqrt{4}}} + \frac{\sqrt{4}}{.\overline{4}} = 113$$

The pastime was first mentioned in the issue for December 30—in the palindromic, invertible year 1881—of a lively London weekly that had been founded that year by the astronomer Richard Anthony Proctor. He called his periodical *Knowledge: An Illustrated Magazine of Science, Plainly Worded—Exactly Described.* A letter to the editor expressed astonishment at the fact (shown to the writer by a friend) that all integers from 1 through 20, except 19, could be expressed by four 4's and simple signs. Factorials and dots were not allowed. Readers were asked to try their hand at it before solutions were given in a later (Jan. 13) issue. (With the help of the factorial sign, 19 can be expressed: $4! - 4 - 4/4$. Can the reader of this book find a way to do it by using only the four arithmetical signs and the decimal point? See Answers, Five, II.)

Since 1881 the game has enjoyed occasional revivals. A lengthy article on the topic, by W. W. Rouse Ball, appeared in the *Mathematical Gazette* for May 1912, and there have been scores of subsequent articles, including tables that go above 2,000. Even now the mania will suddenly seize the employees of an office or laboratory, sometimes causing a work stoppage that lasts for days.

"Is it possible," I asked Dr. Matrix, "to express 1964 with four 4's and the traditional symbols?"

$$1 = \frac{44}{44}$$

$$2 = \frac{4}{4} + \frac{4}{4}$$

$$3 = \frac{4+4+4}{4}$$

$$4 = 4(4-4)+4$$

$$5 = \frac{(4\times4)+4}{4}$$

$$6 = 4 + \frac{4+4}{4}$$

$$7 = \frac{44}{4} - 4$$

$$8 = 4+4+4-4$$

$$9 = 4+4+\frac{4}{4}$$

$$10 = \frac{44-4}{4}$$

$$11 = \frac{44}{\sqrt{4}+\sqrt{4}}$$

$$12 = \frac{44+4}{4}$$

$$13 = \frac{44}{4} + \sqrt{4}$$

$$14 = 4+4+4+\sqrt{4}$$

$$15 = \frac{44}{4} + 4$$

$$16 = 4+4+4+4$$

$$17 = (4\times4) + \frac{4}{4}$$

$$18 = (4\times4)+4-\sqrt{4}$$

$$19 =$$

$$20 = (4\times4)+\sqrt{4}+\sqrt{4}$$

Figure 7. The problem of four 4's

He shook his head vigorously. "Of course, many important dates *are* possible. 1776 is 4 times 444. But 1964 is not one of them. With five 4's, yes." He jotted on my note pad:

$$44^{\sqrt{4}} + 4! + 4$$

"But four 4's, no."

"How about 64?"

"That," said Dr. Matrix, "is not difficult. Oddly enough, 64 can also be expressed—under traditional restrictions, of course—with three 4's and also with two."

The reader is invited to try his skill on all three problems; that is, to express 64 with four 4's, with three 4's, and with two 4's. No symbols may be used other than those that have been mentioned. The task is middling hard with four 4's, ridiculously easy with three, extremely difficult with two. (See Answers, Five, III.)

Dr. Matrix gazed vacantly off into space when I spoke to him about the coming election campaign. In an interview that I reported in chapter 2 we had spoken about the grim numerological pattern of death in office for every president who had been elected in a year ending in 0, beginning with Harrison's death in 1841. Now Kennedy, who had been elected in 1960, had been killed by an assassin.

"Yes," he said finally, "the names and birth dates of the leading candidates deserve careful analysis. In the past twenty-two elections, beginning in 1876, the only occasion on which the man with the shorter last name won a majority of the popular vote was in 1908, when Taft defeated Bryan. This gives Rockefeller an edge over all his competitors. Of course Nixon, Romney, and Johnson are eliminated because their names lack a double letter such as the two *l*'s in Rockefeller."

I was scribbling furiously. "That makes Rocky a stronger candidate than Goldwater, I suppose. Both men have the double letter, but Rocky's last name is longer."

"In that respect, yes. Rocky's height, of course, is a lia-bility. In the past fifteen elections, beginning in 1904, the only time the shorter candidate won the popular vote was in 1940, when Roosevelt, at six feet two inches, defeated Willkie, six feet two and a half. By the way, did you know that both Rockefeller and Romney, the two R.-initial men, were born on July 8?"

I shook my head.

"In fact, all five leading Republican candidates—Rockefeller, Romney, Goldwater, Nixon, and Scranton—were born in months that begin with J. Goldwater and Nixon were born in January, Scranton in July. Now, j is the tenth letter of the alphabet. Note that REPUBLICAN has ten letters and that the digits of '64 sum to 10."

"Is that a good omen?"

"To a certain degree. The digits of 1964, however, sum to 20. The only candidate with exactly 20 letters in his full name is Barry Morris Goldwater. On the other hand, the president will not be inaugurated until 1965, which sums to 21, the number of letters in the name of William Warren Scranton."

"Your numerology is confusing," I said.

"No more than politics. I regret to report that Scranton, the governor of Pennsylvania, was not born in Scranton, Pennsylvania, or in its anagrammatic cousin, Cranston, Rhode Island. He was born in Madison, Connecticut. But MADISON is a presidential name, so that should be counted a favorable sign."

"Someone has suggested," I remarked, "that Rockefeller should open a campaign speech by saying: 'I come to Barry Goldwater, not to praise him.' "

Dr. Matrix looked as solemn as an owl. "It is possible to devise many appropriate puns on the candidates' names. Nix on Nixon, for example. Aldrich Rockefeller sounds like 'old rich rocky feller,' and one might say that his views on certain issues are enough to rock a feller. Nixon's straightforward Republicanism is indicated by the fact that the first and last letters of REPUBLICAN are his initials; the same letters backward are Rockefeller's. The New York governor's full initials, backward, may be prophetic: the Republican who *RAN* in 1964. Crossword puzzle experts will recognize that Roc/kef/ell/er are respectively the words for a fabled bird, dreamy tranquillity, a unit of length, and an Irish god of the sea."

"Interesting," I said, "but all much too ambiguous. Can't you give me a specific prediction about the Republican nomination?"

"My guess is Barry Goldwater. The contest was foreshadowed in 1844 by the publication in England of William Makepeace Thackeray's picaresque novel *The Luck of Barry Lyndon*. Barry Lyndon, in case you don't know the book, is an Irish rascal—soldier, gambler, politician, braggart, and fortune hunter. No matter how shameful his actions, or how much he's criticized, he never doubts that he's the finest and wisest of men."

Iva, who had left us to ourselves while we conversed, returned to the living room to remind her father that he was due back at the Purple Hat in half an hour. I paid the taxi fare and walked them around to the club's back entrance.

"When is the last show over?" I asked Iva.

"Two thirty," she said, smiling.

"Will you, perhaps, feel like eating somewhere?"

"I'll be ravenous."

How does a man in middle age kill three hours in the middle of a freezing, snowy night, in the middle of the week, in mid-December, in the middle of the Windy City? I decided to return to my hotel and spend the time in the middle of my bed.

6. Miami Beach

Dr. Matrix's appearance at the Purple Hat Club, in Chicago, was so sensationally successful that he obtained a six-week booking at one of the plushier Las Vegas hotels. According to several well-authenticated reports, the old swindler not only got a top price for his act but also managed to win $70,000 at the blackjack tables by playing his own modification of the system explained by mathematician Edward O. Thorp in his eye-opening book *Beat the Dealer*. It is said that Miss Toshiyori helped her father by making surreptitious calculations on a small transistorized computer concealed in her handbag.

I wrote to Dr. Matrix in September 1964, while he was still in Vegas, asking for his opinions on the coming presidential contest between Goldwater and Johnson. I reminded him that only Barry Goldwater had the valuable double letter in his name. Did this mean he would win?

No, Matrix replied. The double letter was balanced by another important rule. Voters tend to prefer names ending in *on* over names ending in *er*. Tyler, Hoover, and Eisenhower were the only *"er"* presidents, as compared with nine *"on"* presidents: Washington, Jefferson, Madison, Jackson, William Henry Harrison, Andrew Johnson, Benjamin Harrison, Wilson, and Lyndon Johnson. To be sure, Eisenhow*er* did defeat Stevens*on*, but that was a rare exception. Lynd*on* Johns*on* had a double *on*. That was enough, said Dr. Matrix, to overcome the power of the double *r* in Barry.

After a careful study of the many complex numerological aspects of the contest, Dr. Matrix constructed a number—13,212—that he told me concealed an absolutely infallible prediction of the name of the next president. So certain was he of this that he asked me to publish the prediction in my October column in *Scientific American* (which I did), along with the statement that he would send $100 to every magazine subscriber if his prediction proved to be wrong. He did not wish to influence the election, he wrote, by decoding the prediction before November 3. He promised to send me a full explanation in time for publication in my December column.

After closing their run in Vegas, Dr. Matrix and Iva decided to spend some of their newly acquired loot on a skiing vacation in western Canada. In mid-October I received a postcard from Iva that was postmarked Zero, Montana. They had bought a Jaguar in Canada, she said, and were now planning to drive leisurely around the country on their way to Miami, Florida.

When Iva's second card came from Unityville, South Dakota, I guessed that the pair were on a winding route that would probably take them through towns with numerical names in ascending serial order. Cards followed from Two Rivers, Wisconsin; Triplet, Virginia; Four Oaks, North Carolina; Five Forks, West Virginia; Six, West Virginia; Seven Mile, Ohio; Eight Mile, Alabama; Nine Point Mesa, Texas; and Ten Mile, Tennessee.

The next communication, a telegram from Dr. Matrix, arrived on November 4, the day after the election. This is how the doctor decoded his numerical prediction. First the number is partitioned 13/21/2. These three numbers, taken in reverse order, tell us to check the second, twenty-first, and thirteenth letters from the *end* of the Pledge of Allegiance to the Flag. The three letters, in order, are *l, b,* and *j,* the initials of Lyndon Baines Johnson. Dr. Matrix added that he and his daughter were staying at the Moral Rot Hotel in Miami Beach. Would I care, he asked, to pop down for a visit.

I would indeed. After consulting an atlas I was tempted to drive down, sending them postcards from Odd, Virginia, and Evensville, Tennessee, but on second thought it did not seem worth the trouble. I took a plane.

At three that afternoon I found Dr. Matrix and Iva sitting in the hotel's Marquis de Sade Cocktail Room. Money had agreed with both of them. Dr. Matrix had put on weight. His bony cheeks had filled out, so that he now resembled less a green-eyed hawk than a green-eyed owl. In the lounge's dim light Iva looked younger and more tantalizing than ever.

"It's good to have more evidence you're not infallible," I said to Dr. Matrix. "The last time we met you added up dates in such a way as to suggest that a third world war would begin this year."

"I did nothing of the kind," snapped Matrix, looking a trifle annoyed. "I merely suggested that 1964 was a dangerous year in view of certain numerological patterns. Don't be too optimistic. It was last August, remember, that United States planes first began to bomb the bases in North Vietnam. That was a major escalation of the war and one that surely brought us closer to the brink."

"He was right about Johnson's election last month," said Iva.

"Yes, my dear. But your father's prediction was rather ambiguous and left much to be desired. It's true that it

yields *L.B.J.* when applied to the Pledge of Allegiance, but
one of my readers has just pointed out that if you partition
the numeral as 13/2/12, then consult the thirteenth chapter
of the King James Bible—that is, the thirteenth chapter of
Genesis—verse 2, you'll find that the twelfth word is
GOLD, a clear allusion to Barry Goldwater."

Dr. Matrix blinked his eyes solemnly and Iva smiled
faintly. "An astonishing coincidence," he said, "but I'm really not surprised. Improbabilities, you know, are extremely probable, and they always mean something. Did I
ever mention that in the Dewey decimal system for li-
braries the classification number for numerology is
133.335?"

I took pencil and paper from my jacket as I shook my
head.

"If you add that number to its reversal, 533.331," Dr.
Matrix continued, "the result is 666.666—a double form of
the mark of the Beast."

"That interests me," I said. "I've just had a book published called *The Ambidextrous Universe*. It's all about
mirror reversals and left-right symmetries."

"I'm halfway through it. An amusing book, but I wish
you had consulted me before you wrote that section on
words, letters, and numbers. I could have given you much
better material."

"Examples, please," I said, pencil poised.

Iva glanced at her wristwatch. "You boys must excuse
me. I want to get in a swim before the sun's too low." She
nodded in my direction. "See you at dinner, Nitram."

"Consider," Dr. Matrix went on after we had reseated
ourselves, "the present international scene. Surely the
great left-right split between the United States and the So-
viet Union is mirrored by the fact that the initials of the
two giants are left-right reflections: *U.S.* and *S.U.* And have

you noticed that the famous *K.* and *B.* team of Khrush and Bulge that followed Stalin is echoed by the new *B.* and *K.* team of Brezh and Kos? Only now the order is reversed— the *B.*, not the *K.*, is on top."

Dr. Matrix borrowed my pencil and rapidly jotted down the value of pi to thirty-two decimals (see Figure 8). "Mathematicians consider the decimal expansion of pi a random series, but to a modern numerologist it's rich with remarkable patterns."

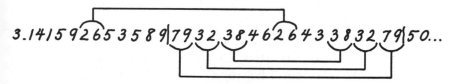

Figure 8.   A curious reflection pattern in pi

He bracketed the two appearances of 26. "You observe that 26 is the first two-digit number to repeat. Note how the second 26 marks the center of a bilaterally symmetrical series." Dr. Matrix inserted vertical bars to enclose eighteen numerals, then bracketed six other number pairs as shown. The number pairs 79, 32, and 38 on the left are balanced by the same three pairs in reverse order on the right! He also called attention to the sets of five digits on each side of the first 26. The first set sums to 20, the number of decimals preceding the second 26. The second set sums to 30, the number of decimals preceding the second bar. Together they sum to 50, the two-digit number following the last bar. The sequence between bars starts with the thirteenth decimal and 13 is half of 26. The three pairs—79, 32, and 38—have six digits that sum to 32, the pair in the middle as well as the total number of decimals shown. The 46 and 43 on each side of the second 26 sum to 89, the number pair preceding the first bar.

"I could talk for hours about that crucial number 32," Dr. Matrix continued. "It's one of the ubiquitous constants of nature. An object falling to the earth accelerates 32 feet per second per second. Water freezes at 32 degrees Fahrenheit. There are 32 crystal classes. Man has 32 teeth. There are 32 electrons in the filled fourth energy level of atoms. There are 32 fundamental, long-lived particles. Eddington's fine-structure constant, 137, is the 32d odd prime. And so on. Of course 32 is 2 raised to the power obtained by adding 3 and 2."

"It's too bad that the number of moons in the solar system is only thirty-one. How do you account for that?"

"My dear Gardner," he replied. "There are thirty-*two* moons in the solar system. There is one moon that our astronomers have not yet discovered."

"Could you tell me what planet it belongs to?"

"3 times 2 is 6. The sixth planet from the sun is Saturn." *

"The science writer Guy Murchie recently pointed out to me," I said, "that Franklin D. Roosevelt, the 32d president of the United States, was elected in the year '32. And I seem to recall there are 32 counties in the Irish Free State. Wasn't that why James Joyce, in *Finnegans Wake*, used 32 as a symbol for the fall of Finnegan?"

---

* This prediction was confirmed in 1967 when Audouin Dollfus, of the Meudon Observatory, France, reported the discovery of a new moon (later named Janus) very close to the outer edge of Saturn's rings. (Unfortunately, a 33d moon, a tiny satellite of Jupiter, was discovered in 1974.) For a charming numerological theory in which the numbers of satellites around each planet are deduced from the structure of energy levels in the atom, consult Sam Elton, *A New Model of the Solar System* (New York: Philosophical Library, 1966), chapter 5.

Dr. Matrix nodded. "Someday I intend to write a commentary on Joyce's number symbolism. But back to pi. I wish I had time to go into the subtler properties and the historical significance of those first 32 decimals. Let me say, though, that '62–'64 in the center of that series between the bars indicates the three eventful years that have just passed as the world moves from '33 on the right, the year Hitler became chancellor, to George Orwell's '84 on the left. Correctly interpreted, you know, pi conveys the entire history of the human race."

"Do you have," I asked, shifting the topic purposely, "any left-right reversal items that might provide puzzles?"

"Thousands," sighed Dr. Matrix. "Consider for a moment the digits from 1 to 9. Arrange them in descending order, reverse and subtract (see Figure 9). The result is quite unexpected. The same nine digits reappear in the answer."

"I've seen that before," I said, "in medieval books on numerology."

"Of course," replied Dr. Matrix. "I bring it up because

$$
\begin{array}{r}
987654321 \\
-123456789 \\
\hline
864197532
\end{array}
\qquad
\begin{array}{r}
9876543210 \\
-0123456789 \\
\hline
9753086421
\end{array}
$$

$$
\begin{array}{r}
98754210 \\
-01245789 \\
\hline
97508421
\end{array}
\qquad
\begin{array}{r}
954 \\
-459 \\
\hline
495
\end{array}
$$

Figure 9.   Four sets of distinct digits that are self-replicating

few recreational mathematicians know the other examples of what I call 'self-replicating sets' of distinct digits."

The problem, as Dr. Matrix explained in more detail than I can give here, is that of finding a set of n digits, no two alike, such that when they are arranged in descending order and reversed and the new number is subtracted from the old, the same n digits reappear in the result. No examples exist, Dr. Matrix assured me, for sets of one, two, five, six, or seven digits (except for the trivial case of $0 - 0 = 0$). For three digits the unique example is 954. For eight digits 98,754,210 is unique. The example of nine digits given above is also unique, and for ten digits 9,876,543,210 is obviously the only case. For four digits there is again only one example. Can the reader discover it? (See Answers, Six, I.)

"By the way," Dr. Matrix added, "your readers might enjoy *dividing* 987,654,321 by 123,456,789. It's hard to believe, but the answer is 8.00000007+, seven decimal 0's followed by 7.* A pity the quotient isn't exactly 8, but that's the way things are sometimes, in numerology as well as physics. Let's go up to my suite. It's so dark here I can hardly see what I'm writing."

Dr. Matrix paid the tab with a generous tip, and we took an elevator to his rooms on the top floor. "What's *your* room number?" he asked as we made ourselves comfort-

---

* Some readers may wish to probe the whys of this curious division. Carried to fifteen decimal places the quotient is 8.000000072900000+. Is it accidental that 729 is the cube of 9?

The answer is no. Before explaining why, let's first extend the calculation:

8.000000072900000663390006036849054935326399911470239 . . .

Richard H. Hart has observed: (1) The above number begins with a sequence of 7, 5, 3, and 1 zeros alternating with sequences of 1, 3, 5, and 7 nonzero digits except for "the exasperating 0 in the twenty-fifth place after the decimal"; (2) if you transpose the last two digits of

able. After I told him he closed his eyes for a minute, then opened them suddenly and said: "A most unusual number. It's the only three-digit number with the following property: multiply it by a certain digit and the result is a three-digit number that is the reverse of what you get if you add that same digit to it instead." (See Answers, Six, II.)

"I'm writing all this up for January," I said. "I'll include that and give the answer in February. Any puzzles involving the number of the new year?"

"I assumed you'd ask me that," Dr. Matrix answered with one of his rare, crooked smiles. "You recall that in your column for October 1962 you asked readers to insert plus or minus signs wherever they pleased inside the series 123456789 and in the reverse series 987654321 so that in each case the series totaled 100?"

I nodded. "And in the 'Letters' department for January 1963 we printed computer results giving eleven different solutions for the ascending series, fifteen for the descending." (See Figure 10.)

"To complete the record," Dr. Matrix said, "if a minus sign is allowed in front of the first digit, there are three more answers for the descending series and one unique answer for the ascending."

987,654,321, then divide the resulting 987,654,312 by 8, the result is exactly 123,456,789.

Two readers, Fitch Cheney and Alan B. Lees, independently called attention to the following equalities:

$$729 = 9^3 \times 91^0$$
$$66,339 = 9^3 \times 91^1$$
$$6,036,849 = 9^3 \times 91^2$$

This suggested to both men the following conjecture:

$$\frac{987,654,321}{123,456,789} = 8 + 729 \times 10^{-10} \sum_{n=0}^{\infty} (91 \times 10^{-10})^n$$

which they easily verified by using the standard formula for sums of geometric progressions.

SOLUTIONS FOR ASCENDING SEQUENCE

$$123-45-67+89 \qquad =100$$
$$123+4-5+67-89 \qquad =100$$
$$123+45-67+8-9 \qquad =100$$
$$123-4-5-6-7+8-9 \quad =100$$
$$12-3-4+5-6+7+89 \quad =100$$
$$12+3+4+5-6-7+89 \quad =100$$
$$1+23-4+5+6+78-9 \quad =100$$
$$1+2+34-5+67-8+9 \quad =100$$
$$12+3-4+5+67+8+9 \quad =100$$
$$1+23-4+56+7+8+9 \quad =100$$
$$1+2+3-4+5+6+78+9=100$$

SOLUTIONS FOR DESCENDING SEQUENCE

$$98-76+54+3+21 \qquad =100$$
$$9-8+76+54-32+1 \qquad =100$$
$$98-7-6-5-4+3+21 \quad =100$$
$$9-8+7+65-4+32-1 \quad =100$$
$$9-8+76-5+4+3+21 \quad =100$$
$$98-7+6+5+4-3-2-1=100$$
$$98+7-6+5-4+3-2-1=100$$
$$98+7+6-5-4-3+2-1=100$$
$$98+7-6+5-4-3+2+1=100$$
$$98-7+6+5-4+3-2+1=100$$
$$98-7+6-5+4+3+2-1=100$$
$$98+7-6-5+4+3-2+1=100$$
$$98-7-6+5+4+3+2+1=100$$
$$9+8+76+5+4-3+2-1=100$$
$$9+8+76+5-4+3+2+1=100$$

Figure 10.   A computer's solution of two problems with the nine digits

For the interested reader these are: $-9 + 8 + 76 + 5 - 4 + 3 + 21;\ -9 + 8 + 7 + 65 - 4 + 32 + 1;\ -9 - 8 + 76 - 5 + 43 + 2 + 1;\ -1 + 2 - 3 + 4 + 5 + 6 + 78 + 9$.

"You might ask your readers to see if they can insert five signs within the ascending series to make a total of 65, the new year. Five is the smallest number of signs that will do it, and there's only one answer, even if a minus sign is permitted in front of the 1."

"How about the descending series?"

"If no minus sign is allowed in front, the minimum number of signs required to make 65 is six, with five different solutions. They're of no special interest. But if a minus sign in front of the 9 *is* permitted, there's a unique five-sign solution. Your readers might enjoy looking for that also." (See Answers, Six, III.)

"I'm sure many will," I said. "But you must admit that all the problems you've mentioned so far are relatively trivial. Can you give me something with a bit more substance?"

Dr. Matrix stood up, walked over to a desk and returned with what appeared to be a silver ruler. When I examined it, I saw that it bore only four marks (see Figure 11). "An old friend in Tokyo sent me this," said Dr. Matrix. "It's thirteen inches long and the marks are placed so that one can measure exactly any integral length of inches from 1 through 13."

I studied the ruler. "I see what you mean. From 0 to 1 measures one inch. From 0 to 2 measures two. Three can be measured by the 10 and 13 marks, four by the 6 and 10 marks, and so on."

Three marks, Dr. Matrix told me, are sufficient for a nine-inch rod if one wishes to measure, in one step, any integral length from one through nine inches. But on a twelve-inch ruler four marks are required for measuring

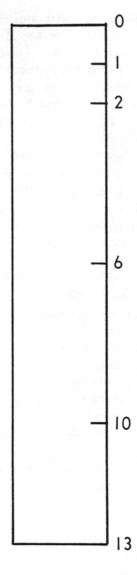

Figure 11.   Dr. Matrix's thirteen-inch ruler

lengths from one through twelve inches. Can the reader find a suitable way of placing the four marks on a twelve-inch rod, and prove that three are not enough? And what is the maximum number of different integral lengths that three marks, suitably placed on a twelve-inch rod, *will* measure?

A more difficult question: On a yardstick thirty-six inches long, what is the smallest number of marks that permit measuring all integral lengths from 1 through 36? Show how the marks should be placed. (See Answers, Six, IV; the general problem of finding a formula or procedure for obtaining the minimal markings of rods of any length is still unsolved.)

Dr. Matrix and I were discussing various combinatorial approaches to the ruler problem when Iva came into the room wearing a bright orange bikini. I stopped thinking about combinatorial arithmetic to concentrate my attention on space curves and the dynamics of oscillating solids.

7. Philadelphia

Dr. Matrix and his daughter, I had discovered over the years, seldom remained long at one address. I was beginning to suspect that their peripatetic life was designed as much to avoid the police as to escape local creditors. For ten months after my last visit with them, in Miami Beach, I lost all track of them. Then, early in November 1965 I received from Iva a cryptic postcard. Its entire message was:

$$A, B, ...$$
$$. . . . .$$
$$. . . . .$$
$$. . . . .$$
$$. . . . .$$
$$. . . . .$$

The card was postmarked in Philadelphia, and Iva had added a phone number. I puzzled over the message for several days before the meaning suddenly came to me: "Long time no see."

Two days later I was standing outside the door of an office in an old building in Philly, inspecting the brass shingle above the bell. It read: "Dr. Irving J. Matrix, psycho-numeranalyst."

Iva's smile, when I stepped into the anteroom, was as dazzling as ever. But before we had had time for more than preliminary pleasantries she escorted me into Dr. Matrix's large, dimly lighted sanctum, then quietly disappeared. A tall figure rose from behind an enormous marble-topped desk.

"Good to see you again, Gardner," Dr. Matrix said, with a thick, almost convincing Viennese accent. Since I had last seen him he had also acquired a small goatee, reddish brown and pointed like a triangle. A pince-nez was clipped to his hawklike nose, and through its lenses (probably plain glass) his emerald eyes glittered with inscrutable humor.

Did he seriously believe in the principles of his new profession? If not, he gave no indication; indeed, there was a curious ring of plausibility about the techniques he explained. He was working, he said, in the analytic tradition of Freud, but combining it with a much stronger emphasis on the unconscious mind's awareness of the symbolic properties of numbers and letters, and with therapeutic methods borrowed from Russian psychiatry. Freud had been on the right track, he maintained, in those early years when he had taken seriously the numerological theories of his good friend Wilhelm Fliess.*

"Unfortunately," said Dr. Matrix, adjusting his pince-nez, "Freud was, by his own admission, a poor mathematician, and so his attempts at numerological analysis tended

---

* Fliess was a Berlin nose doctor who became obsessed by the numbers 23 and 28; they were the basis for a theory of cycles that he applied to everything from the nose to the solar system. For about a decade he and Freud were involved in a strange neurotic friendship. I have given an account of Fliess's numerology, and its modifications by contemporary followers, in chapter 12 of my *Mathematical Carnival* (New York: Knopf, 1975).

to be either trivial or absurd. Do you recall his explanation of the number 2,467?"

I nodded. Freud had written Fliess that he had finished checking the page proofs of his *Interpretation of Dreams* and would make no more corrections even if they contained 2,467 mistakes. Before mailing the letter Freud asked himself: Why did that particular number, seemingly random, pop into his skull? "Nothing that happens in the mind is arbitrary or undetermined," he wrote in the letter's famous postscript. He then proceeded to give what he believed were the unconscious determinants of 2,467, an analysis he later added to the section on numbers in the last chapter of his *Psychopathology of Everyday Life.*

"Freud's explanation of 2,467," I said, "always seemed farfetched."

"It is completely unconvincing," agreed Dr. Matrix. "If Freud had known something about number theory, he might have recognized 2,467 as a prime. He had just finished writing his greatest book. It was the *prime* year of his life. A year contains 365 days, so what could be more natural than to seize on the 365th prime, which is 2,467?" *

"But," I said incredulously as I jotted this in my notebook, "if Freud was so poor in mathematics, how could his unconscious mind have picked out the 365th prime?"

"You forget, my dear Gardner, the *collective* unconscious, so brilliantly revealed in the work of Jung. Primes,

---

* It is perhaps worth mentioning that the 367th prime, 2,477, was dreamed by one of Jung's patients. Jung's absurd analysis of the meaning of this number (in "On the Significance of Number Dreams," *Collected Papers on Analytical Psychology*, 2d ed. Baillière, Tindall and Cox, 1917, pp. 190–199) involves the birth dates of the patient, his wife, his mistress, his mother, and his two children, as well as his age and the age of his mistress, and a few other things.

the building blocks of the integers, are deeply etched in the collective memory of the human species. The power of our unconscious to manipulate numbers and other symbols is far greater than even Jung and his disciples dared imagine."

Dr. Matrix reached for a book on his desk—a copy of *The Scientist Speculates: An Anthology of Partly-Baked Ideas* (New York: Basic Books, 1962)—and opened it to page 331. On that page I. J. Good, a mathematician at Trinity College, Oxford, wonders why space has three dimensions. "We may try to run away from the question," Good writes in a passage Dr. Matrix had marked, "by saying that 3 is a small enough number not to need an explanation. An explanation would have been more in demand if the dimensionality had been 32,650,494,425."

"Why," Dr. Matrix asked, leaning back and removing his pince-nez, "did *that* number come into Good's conscious mind?"

I chuckled when he showed me how simply Good's number could be obtained. (The reader is invited to try his analytical skill on this—thirty minutes is par—before he turns to Answers, Seven, I.)

One of Dr. Matrix's diagnostic techniques, he explained, is to have the patient lie comfortably on his back, in a dark room, and free-associate while single numerals are projected in colored light on the ceiling: "I had a patient last week," he told me, "who became unusually agitated whenever a green 4 appeared. It turned out that he was stealing money—green stuff—from the cash register of the store where he worked. His superego was disturbed by 4's symbolic honesty."

"Its *honesty?*"

"Yes. 4 is the only number, among all the infinity of

numbers, that states correctly the number of letters in its English name." *

Another diagnostic test, also invented by Dr. Matrix, consists of giving the patient ten large cards, bearing the numerals 0 through 9, and asking him to arrange them in any order to form a ten-digit number. He cited the case of a woman patient with the unusual name of Aniba Di Figby. Using the familiar code of $a = 1$, $b = 2$, and so on, her names translate to 114,921, 49, and 697,225, all square numbers. Moreover, she boasted the attractive measurements 36-25-36, her height was 64 inches, and she had been born in the square year 1936. The well endowed Miss Figby thought of herself as "hippy," but obviously was a "square." Dr. Matrix was therefore not surprised when she arranged the ten cards to form 9,814,072,356, the largest square that can be made with the ten digits. Its square root, 99066, is interesting in its own right, Dr. Matrix pointed out: it remains unchanged when inverted.

"Another patient," Dr. Matrix went on, "was a businessman whose chief concern in life was maximizing his firm's profits. His unconscious naturally arranged the ten cards so that, if you made a dividing line between two cards, the cards on the left and right sides of the line formed two numbers that multiplied together to give the largest possible product. You might ask your readers to discover what those two numbers were."

"Excellent," I said as Dr. Matrix showed me a clever way to find the answer. (See Answers, Seven, II.)

"Your readers might also enjoy searching for the unconsciously determined order behind *this* number." Dr. Matrix paused to put on his pince-nez and write 8,549,176,320

---

* For comments on this remark by readers of my *Scientific American* column, see Answers, Seven, closing section.

in my notebook. "The first name of the woman who produced that arrangement of the ten digits is Betty. As another clue, you can add that she suffers from a compulsion to put things in order, a compulsion that has found an outlet in her job as indexer for a large textbook publisher in Manhattan." (See Answers, Seven, III.)

Any number that comes to one in a dream, Dr. Matrix emphasized, is of special significance in psychonumeranalysis, but the analyst must be ingenious and flexible if he is to interpret the number correctly. Dr. Matrix had a low opinion of Freud's dream-number explanations in his book on dreams, and he thought the later attempts recorded by Jung, Adler, Stekel, and Jones were equally humdrum.

"I had a Pentecostal minister in to see me recently," he said, "who repeatedly dreamed of 7,734. When I asked him to write that number and then turn it upside down, he confessed at once that he had been suffering for years from a fear that his religious doubts would deprive him of his place among the saved. His dream censor was, of course, concealing HELL, the feared word, by inverting it. A similar case concerned a Boston financier who told me about his recurring dream of 710. He had lost a huge sum of money by investing in a worthless Texas oil company. The upsetting word OIL was being turned around in his dream to keep him from waking up. I doubt if Freud would have been able to interpret either of those numbers properly."

The name of the city in which one lives also plays a major role in psychonumeranalytic diagnosis. Dr. Matrix told me he had several patients from Pleasantville, in northern Westchester County, New York, who were severely disturbed by the contrast between the town's name and the kind of life they lived there as employees of *Reader's Digest*. The exclusion of Jews from Bronxville, in

southern Westchester, is clearly a Gentile reaction to
BRONX in the town's name. In Philadelphia, he said, the
slightest feelings of hostility toward a brother are mag-
nified by the slogan "City of Brotherly Love."

Street names also reflect unconscious correlations. Can
anyone doubt, he asked, that the high incidence of mental
breakdowns among New York advertising men is as-
sociated with the MAD in Madison Avenue? I myself, he
reminded me, live on Euclid Avenue. And did I know that
William Feller, the probability expert, lives in Princeton
on Random Road?

One's own name and its initials are also basic psychonumeranalytic symbols. It is no accident, said Dr. Matrix,
that Adam Clayton Powell's initials are the last three letters of the NAACP.* Did the fact that James Augustine
Aloysius Joyce's initials form a palindrome have anything
to do with his compulsion toward word play? Did Herbert
Clark Hoover's palindromic initials influence his efforts to
move the country backward? Was Edgar Allan Poe impelled to become one of America's greatest poets by the
fact that a single letter affixed to his name changed it from
Poe to Poet? Did not poetry suit Poe to a *T*?

Dr. Matrix said he had once treated a woman with the
maiden name of Mary Belle Byram. She was always doing
things backward and had not realized, until he called it to
her attention, that her entire name was a palindrome. "She
was married, by the way," Matrix added, "to an army of-
ficer. Army men had always fascinated her. She never suspected that this was because ARMY and MARY are anagrams."

I find in my notes several other startling instances of
similar correlations that Dr. Matrix had encountered clini-

---

* Never Annoy Adam Clayton Powell.

cally. A man named Dennis (SINNED, backward) was tortured by guilt feelings over his relationship with a Russian girl named Natasha (AH, SATAN, backward). A teenaged boy named Stewart wanted to be a baseball player because his name is an anagram of SWATTER. An attractive spinster schoolteacher whose last name was Noyes had turned down many marriage proposals because her name can be partitioned into NO-YES, so that she could never make up her mind. Reversing the YES produces NOSEY, a fact that explained her fondness for malicious gossip about her colleagues. A young man named Andrew had become a drifter, prodded unconsciously by the fact that ANDREW is an anagram of WANDER. A boy named Cyril could not understand his compulsion to compose LYRIC poems. An artist named Bernard was unable to paint a landscape without putting somewhere in it a RED BARN.

Publishers, Dr. Matrix assured me, have an unconscious tendency to sign up authors whose last names are anagrams of the house. He cited the case of Selden Rodman, who edited a poetry anthology for Random House, and Robert Gover, whose books are published by Grove Press. "You can understand," he said, "why Salvador Dali recently chose Dial Press as the publisher of his autobiography."

My notes contain many other instances cited by Dr. Matrix of how the names of prominent people are often unconsciously correlated with behavior patterns. It was no accident, he said, that Hetty Green became the nation's greatest financier (collector of "green stuff") of the nineteenth century; that the neurologist Lord Brain (he died in 1966) was England's most eminent brain specialist; that Jerald Carl Brauer, dean of the University of Chicago's divinity school, has the initials *J.C.*; that the psychiatrist who collaborated with Norman Vincent Peale in founding the minister's psychiatric clinic was the late Smiley Blanton; that Lyndon Johnson greatly admired the art of Norman Rockwell because Rockwell's political views lay between the extremes of Rockwell Kent on the left and George Rockwell on the right; that Norman Mailer's father, I. B. Mailer of Brooklyn, obviously became an accountant because his initials were *I.B.M.*

An amusing episode involved a patient named Smith. He had come to Dr. Matrix for advice on how to cure himself of chronic constipation. Dr. Matrix was so annoyed by this request—he felt that the man should have gone to a medical doctor—that he told Smith to go home and do the following. He was to write down any number of three digits, provided the first and last digits had a difference of at least two. He was then to reverse the number (write the digits in reverse order) and subtract the smaller number from the larger. The result was to be reversed again, but this time the two numbers were to be added. The final sum would be, Matrix said, a four-digit code word that would tell him what he should do. The digits were to be translated by means of the following key which Matrix wrote on a card and gave to the man:

$$
\begin{array}{ccccc}
S & M & I & T & H \\
1 & 2 & 3 & 4 & 5 \\
6 & 7 & 8 & 9 & 0
\end{array}
$$

(Can the reader explain why this works out as it does? See Answers, Seven, IV.)

Dr. Matrix spoke also of the subtle sexual symbolism invariably concealed in the names of the great love goddesses of the screen. Jean Harlow's last name is only one letter away from HARLOT. Ursula Andress's last name is only one letter away from UNDRESS. Among capital letters the two strongest breast symbols are obviously *M* and *B*. It

is no accident, Dr. Matrix said, that the names Marilyn
Monroe and Brigitte Bardot have the initials *M.M.* and
B.B. The letter symbolism is reinforced by the doubling of
the letters. Note also that these letter pairs provide the
consonants of MAMA and BÉBÉ (French for BABY).

I wish I had space to discuss some of Dr. Matrix's
clinical experiences with the hidden meanings of phone
numbers, social-security numbers, car licenses, street
addresses, ZIP codes, and Blue Cross numbers. Nor can I
give details of his conditioned-reflex therapy beyond saying that it involves administering electric shocks to a re-
clining patient whenever certain words and numbers are
flashed on the ceiling and simultaneously recited aloud by
the patient. After a series of thirty-two such shocks there is
a period of relief from shock while words and numbers of
opposite symbolic meaning are projected and recited.*

The last entry in my notes records a remarkable property
of the then president's name. Dr. Matrix said that it had
been discovered recently by his friend Harry Hazard of
Princeton, New Jersey, and I could have the honor of
being the first to publish it. If Johnson's name is written as
a multiplication problem:

$$\begin{array}{r} \text{LYNDON} \\ \text{B} \\ \hline \text{JOHNSON} \end{array}$$

* Readers interested in the movement called "behavioral psycho-
therapy," from which Dr. Matrix borrowed his Pavlovian shock techniques, can consult two nontechnical articles: "Behavioral Psycho-
therapy," by Albert Bandura, *Scientific American* (March 1967); and
"Freudians Are Wrong, the Behaviorists Say—a Neurosis Is 'Just' a
Bad Habit," by Morton M. Hunt, *New York Times Magazine* (June 4,
1967). The bibliography for Bandura's article lists books defending the
new techniques.

there is a unique substitution of digits for letters that gives the problem a correct numerical form. Solving it is another pleasant exercise that I leave to the reader. (See Answers, Seven, V.)

On my way out I stopped at Iva's desk to ask her a series of curious questions that had been suggested to me by Kirby Baker, then a mathematician at Harvard.

"I have a pair of questions," I said, "each to be answered by yes or no. But before I ask them, may I have your promise in advance that you'll answer them truthfully after you've heard them both?"

Iva's lovely black eyes narrowed a trifle, but she seemed amused. "All right. What are the questions?"

"The first is: Will you have dinner with me tonight? The second is: Will your answer to the first question be the same as your answer to *this* question?"

The poor girl was trapped. She couldn't say no to the first question because then the second question couldn't be answered truthfully with either a yes or no. She had to say yes to both questions. But she was a good sport about it, and we enjoyed, in the most unhurried of United States megalopolises, a relaxed and unhurried evening.

A few weeks later a story in the *New York Times* caught my attention. A wealthy Main Line widow in Philadelphia had died, leaving a small bequest of $50,000 to the Psychonumeranalytical Institute. By the time the story was released to the press the money had actually passed into Dr. Matrix's hands, and the *Philadelphia Inquirer* sent a reporter to ask him how the Institute would use the money. The reporter found the offices completely empty. Dr. Matrix, Iva, and the furniture had vanished without a trace.

8. Pi

After Dr. Matrix and Iva left Philadelphia, and before I saw them again at Wordsmith College, I received a letter from him. It contained a curious prediction about the millionth digit of pi. When I expanded a column on pi for the book collection *New Mathematical Diversions from Scientific American* (New York: Simon and Schuster, 1966), I added a paragraph that summarized Dr. Matrix's letter as follows:

> It will probably not be long until pi is known to a million decimals. In anticipation of this, Dr. Matrix, the famous numerologist, has sent me a letter asking that I put on record his prediction that the millionth digit of pi will be found to be 5. His calculation is based on the third book of the King James Bible, chapter 14, verse 16 (it mentions the number 7, and the seventh word has five letters), combined with some obscure calculations involving Euler's constant and the transcendental number *e*.

Note that Dr. Matrix did not predict the millionth *decimal* digit. His use of the third book of the King James Bible, Leviticus, clearly indicates that he is including the initial 3 in his count. For the dramatic verification of this prediction, see chapter 18.

Figure 12. "I Saw the Figure Five in Gold," Charles Henry Demuth (Metropolitan Museum of Art, the Alfred Stieglitz Collection, 1949)

9. Wordsmith College

For almost a year after Dr. Matrix and his daughter Iva had disappeared from Philadelphia with a $50,000 donation to Dr. Matrix's Psychonumeranalytical Institute, neither I nor the police had been able to discover their whereabouts. Iva usually communicates with me in some cryptic way after a suitable lapse of time, however, so I remained on the alert for a message. One day I received a printed announcement of twelve public lectures on "Combinatorial Aspects of English and American Literature," to be given every Friday evening in Shade Auditorium at Wordsmith College in New Wye, New York. Professor T. Ignatius Marx of the mathematics department was the speaker. The admission charge was $3 a lecture or $25 for the entire series. The weekly topics were listed as follows:

1. The Acrostic Poem.
2. The Palindromic Poem.
3. Concealed and Accidental Verse.
4. Macaronic and Patchwork Verse.
5. Nonsense Verse from *Jabberwocky* to Gertrude Stein.
6. Lipograms, Anagrams, Pangrams.
7. Misplaced Commas.
8. Keys to the Nomenclatures of *Gulliver's Travels,* the *Oz* Books, and Other Descriptions of Imaginary Lands.

9. Lawrence Durrell and the number 4.

10. Vladimir Nabokov and the "Stinky Pinky."

11. Deciphering the Ten Thunderclaps of *Finnegans Wake*.

12. Poetry Programming for Digital Computers.

Who, I wondered, was T. Ignatius Marx? Suddenly it came to me: T. I. MARX is an anagram of MATRIX. He had probably counterfeited credentials as a mathematician and the Wordsmith College administrators had failed to penetrate the disguise.

The announcement arrived by special delivery late Friday afternoon, the day of the first lecture. I dropped everything and drove north to New Wye, in the lower Catskills. It was a good thing I arrived early; the seats in Shade Auditorium were rapidly filling to capacity. Marx was Dr. Matrix all right, although when he first emerged from the wings I did not immediately recognize him. The reddish-brown goatee he had sported in Philadelphia was gone. His prominent nose now overhung a brown handlebar mustache. He was obviously wearing contact lenses, because his green eyes had been transformed to bright blue.

"In the beginning," he started, reading his lecture in a clipped British accent, "was the word. What is a word? It is a combination of sounds that advanced cultures symbolize by combinations of letters. What is a poem? It is a combination of words, chosen not only for their semantic referents but also for their melodic patterns. As Ramón Lull so clearly recognized in the thirteenth century, the poet, like the artist and the musician, is an expert in combinatorics. What is the rhyming dictionary if not a Lullian combinatorial device? With or without such mechanical aids, the poet must explore possible combinations of words until he finds a pattern that maximizes aesthetic satisfactions. A good poem, like a magic square or a crossword puzzle, is an exercise in the great Lullian art of combinatorial thinking. A dictionary is like a box containing thousands of pieces of glass of different sizes, shapes, and colors. As an American poet, Jack Luzzatto, has written:

> In orderly disorder they
>   Wait coldly columned, dead, prosaic—
> Poet, breathe on them and pray
>   They burn with life in your mosaic.*

In addition to maximizing aesthetic values, Dr. Matrix continued, the poet can fashion other remarkable kinds of combinatorial patterns. The acrostic, in which initial letters of lines are in a meaningful order, is perhaps the oldest form of what Dr. Matrix called "meta-aesthetical verse play." The earliest crude examples, he asserted, are found in the Old Testament, where nine Psalms are "abecedarian acrostics," the initial letters of each stanza consisting of the Hebrew letters in alphabetical order.† The first four of the five poems that make up the Book of Lamentations, Dr. Matrix said, are also acrostics of this type, as well as the poem in Proverbs 31, verses 10 through 31, which lists the virtues of the good wife.

Dr. Matrix picked up a pointer that had been leaning against the lectern and tapped it on the floor. Before the

---

* This is the last stanza of a two-stanza poem, "Dictionary," that appeared in the *New York Times* (July 25, 1958).
† In *Finnegans Wake* James Joyce is often, to use one of his own words, "abcedminded." In two places he goes through the entire alphabet: "Ada, Bett, Celia . . . Xenia, Yva, Zulma," and "apple, bacchante, custard . . . xray, yesplease, zaza." There are dozens of instances in which he goes part way, such as "Arty, Bert or possibly Charley Chance," and "Arm bird colour defdum ethnic fort perhaps?" See Bernard Benstock's *Joyce-Again's Wake* (Seattle: University of Washington Press, 1965), p. 28, and footnote on pp. 19–20.

lights dimmed I looked quickly behind me to see who was operating the projector that had been set up in the middle of the center aisle. Yes, it was Iva. Her striking Eurasian features were entirely without makeup. She was obviously masquerading as a student: her dark hair hung below her shoulders and she wore a gray sweater, red boots, and a tweed miniskirt that revealed a splendid pair of knees. I learned later that she was enrolled for undergraduate courses in mathematics and music.

When the room was dark, a picture appeared on the screen of the first stanza of the famous 119th Psalm as it looks in the original Hebrew. Dr. Matrix called attention with the pointer to the fact that each of the stanza's eight verses begins with the letter aleph. The second stanza of the Psalm was then flashed on the screen to show that each of its eight verses begins with beth. The eight verses of the third stanza each start with gimel, and so on through the 22 letters of the old Hebrew alphabet to tau, the last letter.*

The lights went on. It was the Greeks and the Romans, Dr. Matrix declared, who introduced the word and sentence acrostic. He cited numerous instances and showed a slide of each. The prophetic verses of the Greek Sibyls— old ladies who spouted hexameters while in a state of pre-

---

* The best known English abecedarian poem is *The Siege of Belgrade*, which begins:

> An Austrian army, awfully arrayed,
> Boldly by battery besieged Belgrade;
> Cossack commanders cannonading come,
> Dealing destruction's devastating doom;

For this and other examples of English abecedarian verse see William S. Walsh, *Handy-Book of Literary Curiosities* (Philadelphia: Lippincott, 1904), pp. 38–40; and Charles Carroll Bombaugh, *Oddities and Curiosities of Words and Literature* (New York: Dover, 1961), pp. 34–37. Both books may also be consulted on acrostics in general.

tended religious frenzy—were often acrostics. Cicero, in his book *On Divination,* argued that these acrostic features proved that the Sibylline verses were not uttered spontaneously but were carefully composed in advance. The most famous of such acrostics, Dr. Matrix said, was attributed to the Erythraean Sibyl, believed by many scholars to be the same Sibyl who, in book 6 of *The Aeneid,* leads Aeneas into the underworld. The Greek lines were shown on the screen alongside a Latin translation provided by St. Augustine in book 18 of *The City of God.* With the pointer Dr. Matrix showed how the initial letters of the Greek lines formed the five words.

'Ιησους Χριστος Θεου Υιος Σωτηρ

which translate as "Jesus Christ, the Son of God, the Savior."

But there is more. The five initials of the five Greek words form a second acrostic, ιχθυς, the Greek word ICHTHUS, meaning FISH. This explains, said Dr. Matrix, why the fish became such a popular early symbol of Christ. It was often carved on monuments in the Roman catacombs and was widely used as a religious symbol in medieval paintings. It is an appropriate symbol, Augustine wrote, because "Jesus was able to live, that is, exist, without sin in the abyss of this mortality as in the depth of waters."

In a rapid series of slides Dr. Matrix showed many examples of acrostics from the medieval and Renaissance periods, including selections from the fifty acrostic cantos written by Giovanni Boccaccio. He also showed some of the twenty-six graceful acrostics (each on the words ELIZABETH REGINA) in *Hymns to Astraea,* a 1599 book by the English philosophical poet Sir John Davies, and some of the 420 acrostics on names of famous people in Mary

Frege's 1637 work *Fame's Roule*. There were a number of baroque specimens by minor Elizabethan poets, some with the acrostic letters in reverse order and some with a name running down the middle of the poem as well as down both sides.

After the Elizabethan period, Dr. Matrix continued, the English acrostic fell into disrepute. Joseph Addison, writing in volume 60 of *The Spectator,* was unable to decide who was the greater blockhead, the inventor of the acrostic or the inventor of the anagram. Samuel Butler, in his "Character of a Small Poet," described the acrostic writer as one who would "lay the outside of his verses even, like a bricklayer, by a line of rhyme and acrostic, and fill the middle with rubbish." Among the many acrostics by the romantic poets, Dr. Matrix thought the best was John Keats's lyric on the name of his sister-in-law Georgiana Augusta Keats:

GIVE me your patience, sister, while I frame
Exact in capitals your golden name;
Or sue the fair Apollo and he will
Rouse from his heavy slumber and instil
Great love in me for thee and Poesy.
Imagine not that greatest mastery
And kingdom over all the realms of verse,
Nears more to heaven in aught, than when we nurse
And surety give to love and brotherhood.
Anthropophagi in Othello's mood;
Ulysses stormed, and his enchanted belt
Glow with the Muse, but they are never felt
Unbosomed so and so eternal made,
Such tender incense in their laurel shade
To all the regent sisters of the time
As this poor offering to you, sister mine.
Kind sister! aye, this third name says you are;

Enchanted has it been the Lord knows when;
And may it taste to you like good old wine,
Take you to real happiness and give
Sons, daughters, and a home like honeyed hive.

Lewis Carroll was fond of writing acrostics on the names of his young friends. Dr. Matrix displayed several specimens of Carroll's dedicatory verse (for *The Nursery Alice, The Game of Logic,* and *A Tangled Tale*) in which the second letter of each line, instead of the first, spelled the little girl's name. He agreed, however, with my estimate (in *The Annotated Alice*) that Carroll's finest acrostic is the terminal poem of *Through the Looking-Glass,* the first letters of each line spelling the name of the original Alice.

Among American writers he singled out James Branch Cabell as a skillful writer of acrostics, citing the dedicatory poem of *Jurgen* (an acrostic on the name of the critic Burton Rascoe) as a typical Cabellian specimen, as well as the last poem of his *Sonnets from Antan* (1929), which spells "This is nonsense." As instances of the off-color acrostic (I shall not repeat them here) Dr. Matrix exhibited an amusing acrostic attack on Nicholas Murray Butler by the poet Rolfe Humphries (unwittingly printed in *Poetry* magazine under the title "Draft Ode for a Phi Beta Kappa Occasion") and "A Recollection," from page 71 of *The Collected Poems of John Peale Bishop.*

The finest acrostic by an American poet, Dr. Matrix insisted, is the following sonnet:

> "Seldom we find," says Solomon Don Dunce,
>   "Half an idea in the profoundest sonnet.
> Through all the flimsy things we see at once
>   As easily as through a Naples bonnet—
>   Trash of all trash!—how *can* a lady don it?
> Yet heavier far than your Petrarchan stuff—

Owl-downy nonsense that the faintest puff
  Twirls into trunk-paper the while you con it."
And, veritably, Sol is right enough.
The general tuckermanities are arrant
Bubbles—ephemeral and so transparent—
  But *this* is, now—you may depend upon it—
Stable, opaque, immortal—all by dint
Of the dear names that lie concealed within't.

The lady's full name is hidden in an unorthodox way. Can the reader discover it and also identify the poet? (See Answers, Nine, I.)

Dr. Matrix displayed a peculiar acrostic poem which he said had been called to his attention by his old friend Dmitri Borgmann, the word-play expert:

Perhaps the solvers are inclined to hiss,
Curling their nose up at a con like this.
Like some much abler posers I would try
A rare, uncommon puzzle to supply.
A curious acrostic here you see
Rough hewn and inartistic tho' it be;
Still it is well to have it understood,
I could not make it plainer, if I would.

The poem, with the by-line "Maude," appeared in the *Weekly Wisconsin* (Sept. 29, 1888). Can the reader read the concealed words correctly before they are disclosed in the Answers (Nine, II)? The word CON in the second line is a contraction of "contribution."

J. A. Lindon of Addlestone, England, was singled out by Dr. Matrix as the most expert, among living writers of light verse, in the weaving of meta-aesthetic patterns. Lindon's poem "To Those Overseas" was projected on the screen:

A merry Christmas and a happy new year!
Merry, merry carols you'll have sung us;
Christmas remains Christmas even when you are not
   here,
And though afar and lonely, you're among us.
A bond is there, a bond at times near broken.
Happy be Christmas then, when happy, clear,
New heart-warm links are forged, new ties betoken
Year ripe with loving giving birth to year.

It was easy to see that the first line was repeated acrostically by the first words of each capitalized line. Dr. Matrix went on, however, to show that Lindon had inge-
niously worked into his poem a second pattern that is
much more unusual. This pattern too will be explained in
the Answers, Nine, III.

   "For the combinatorial critic," Dr. Matrix continued,
"the unintentional acrostic is even more interesting than
the intentional one. The field is virtually unexplored.
What is the longest word to appear acrostically in Milton's
Paradise Lost? In Pope's *Essay on Man?* In the works of
Yeats, Eliot, Pound, Auden? In the King James Bible?"

   Dr. Matrix displayed several accidental acrostics from
the New Testament. I particularly liked one in which the
three statements of Matthew 7:7 (Dr. Matrix called attention to the triple repetition of 7—twice as a numeral, plus
the seven letters of MATTHEW) are arranged like this:

   Ask, and it shall be given you;
   Seek, and ye shall find;
   Knock, and it shall be opened unto you.

   More remarkable, however, is the following accidental
acrostic Dr. Matrix said he had first learned about when he
checked galleys for *Beyond Language,* by Borgmann (New

York: Scribner, 1967). In Act III, Scene 1 of *A Midsummer Night's Dream* the fairy queen Titania speaks the following lines to Bottom the Weaver:

> Out of this wood do not desire to go:
> Thou shalt remain here, whether thou wilt or no.
> I am a spirit of no common rate;
> The summer still doth tend upon my state;
> ANd I do love thee; therefore, go with me.
> I'll give thee fairies to attend on thee,
> And they shall fetch thee jewels from the deep.

The capital letters on the left spell O TITANIA. "Surely," Dr. Matrix said, "that is the most remarkable unintended acrostic in all English literature. Or was it unintended?"

Dr. Matrix finished with a series of slides displaying unusual examples of acrostics. Of special interest was a photograph of the Dartmouth Street facade of the Boston Public Library, in 1892, showing three memorial tablets on which the following names had been carved:

> Moses
> Cicero
> Kalidasa
> Isocrates
> Milton
>
> Mozart
> Euclid
> Aeschylus
> Dante
>
> Wren
> Herrick
> Irving
> Titian
> Erasmus

When the *Boston Evening Record* discovered that the names formed an acrostic for McKim, Mead, and White, the building's architects, it touched off such a storm of protest that the inscription had to be removed.

The last slide reproduced the final paragraph of Nabokov's short story "The Vane Sisters" from the Winter 1959 issue of the *Hudson Review*. (The story is reprinted in Nabokov's *Tyrants Destroyed,* New York: McGraw-Hill, 1975.) "The Vane Sisters" tells how two dead sisters, Cynthia and Sybil Vane, are influencing, without the narrator's knowledge, the composition of his story about them. An opening descriptive passage mentions icicles and a parking meter. In the story's final paragraph Sybil speaks acrostically as follows:

> I could isolate, consciously, little. Everything seemed blurred, yellow-clouded, yielding nothing tangible. Her inept acrostics, maudlin evasions, theopathies—every recollection formed ripples of mysterious meaning. Everything seemed yellowly blurred, illusive, lost.

The lecture was over. The lights came on, people clapped, and I lost no time in disclosing my presence to Iva. I helped her carry the projector and box of slides to her car, then she drove us to her father's rented house on the outskirts of New Wye. She had assumed the Chinese name Iris Ho Toy, an anagram of her Japanese surname, Toshiyori, and was living in her own apartment near the college. No one suspected, she told me, that Professor Marx was her father.

Dr. Matrix's career at Wordsmith ended abruptly a few weeks after his final lecture. He had announced in that lecture that he had programmed the mathematics department's computer to write modern poetry. All the words in the new *Random House Dictionary of the English Lan-*

guage had been stored in the computer's memory along with rules for combining them in ways derived from an intensive study of the work of ten contemporary poets. The computer had typed out exactly one hundred copies of a long poem, which, suitably bound in imitation leather, could be obtained from Professor Marx for $50 a copy. It was a most impressive poem, although many of its couplets, such as the following, were a bit on the dull side:

> I've measured it from side to side;
> 'Tis three feet long, and two feet wide.

Three weeks later a young instructor in Wordsmith's English department discovered that the poem the computer had typed out was word for word the first version of William Wordsworth's narrative poem "The Thorn." By the time the fraud was discovered, however, Marx and Miss Toy were gone.

10. Squaresville

America is now a healthy country because the squares are taking over, and God bless them.

—Herman Kahn, *Newsweek*,
July 4, 1976, page 30

When the hippie scene started to crash in 1967, and "doing your own thing" went from pad to hearse, thousands of flower children found themselves up tight and no place to glo. Daisy Jones, a minibrained daughter of a friend of mine in Connecticut, finally wandered home. So shaken had she been by the bad vibes around Tompkins Square that she found it impossible to readjust to her former life.

Then Mr. Jones heard about Squaresville.

"Squaresville?" I asked when he told me this in November.

"Yes. It's a hippie rehabilitation center in the northeast corner of Westchester County. Beautiful spot. A psychotherapist named Hawk—Irving J. Hawk—has taken over a new housing development there near Peach Lake."

The "Irving J." made my ears prick up. "Is Dr. Hawk a tall, thin man with a hawk nose and green eyes?"

"Why yes," Jones said. "You know him?"

"I think I do."

"I'd not heard of him before but his therapy seems to work. Daisy went there for nine weeks of 'straightening.' Cost us $1,024 but it was worth every penny. She's back in high school now and making straight A's."

The next morning I drove to Squaresville. Iva seemed surprised and a trifle annoyed when I came into the reception room. "Please," she whispered so that others in the room would not hear, "don't give us away." Then in a normal voice she added, "I'm Mary Jane Grok, the doctor's secretary. Have a seat. Dr. Hawk will see you as soon as possible." *

Iva's black hair was short and neatly trimmed. Large square-shaped red ceramic earrings dangled from her earlobes.

When my turn came to enter Dr. Matrix's office, I found him wearing a gray flannel suit, his face clean-shaven, his gray hair cut in the latest Madison Avenue style. Square links of silver gleamed from his shirt cuffs, and a large square emerald sparkled on a finger ring. On the dark-paneled wall behind him a gold-framed print of Norman Rockwell's portrait of President Johnson looked directly at me with sincere, benevolent eyes. A square poster on the left wall said in large letters: "Be happy, not hippie—learn to play the game." On the opposite wall a similar sign said: "It is better to be rich and healthy than poor and sick."

"I didn't expect you to find us so soon," Dr. Matrix said, pushing a square cigarette box toward me. "Sit down. Have some tar and nicotine."

Squaresville's population, he informed me, was kept at 361, the square of 19, the average age of the patients. When a male hippie arrived, his beard was shaved off and he was given a crew cut. The girls had their hair cut short and coiffured by professional hairdressers. Each hippie

---

* Before the hippie slang of 1967 becomes extinct, perhaps I should record here that marijuana was sometimes called Mary Jane, "grok" meant to dig a scene, and "hawk" was a term for LSD.

was outfitted with conventional clothes, given a wristwatch, and assigned to a room in one of Squaresville's
forty-nine identical split-level houses.

"The nitty-gritty of our treatment," Dr. Matrix went on,
"is intensive conditioning in the art of square living. Each
room has color television. No one is allowed to move a
mattress to the floor or to squat cross-legged. The rooms
are kept supplied with free cigarettes. Everyone has three
square meals a day, including a compulsory martini before
dinner."

"I've been told," I said, "that mathematics is part of
your therapy."

"Yes, an essential part. When these freaked-out young-
sters come to us, they've been flying for months in a
dreamworld. We bring them down to earth by teaching
them there are laws of nature, laws that can be ignored
only at great peril. We show them that they can be as high
as they like but if they jump out a window they won't float
gently down—the law of gravity will kill them. We teach
them there are also laws of health and laws of morality.
The life game, like any other game, is no game at all without rules."

"I still don't see how mathematics is involved."

"I'm getting to that. The hippies, you see, have learned
to overvalue disorder. The random curves of their psyche-
delic art and the whirling patterns inside their circular
mandalas are symbolic of this. Our job is to teach them the
beauties of the straight-sided square. We explain how un-
curved borders make it possible for squares to fit together
without waste space. We show that, as the world population increases, square shapes become necessary for packing into cities, suburbs, subways, commuter trains. In its
interior every square has secret incommensurable quali-

ties, and as many curves as you like can be drawn inside it, but its straight exterior is absolutely essential for close packing."

"I'm beginning to anticipate," I said. "The numerical equivalent of the square is of course the square number."

"Precisely. We begin our conditioning at the lowest levels of arithmetic. First we teach our patients that a square of side n must have an area of n squared. Then slowly, by displaying the elegant properties of square numbers, we convince them that *two* power is much more beautiful than flower power. We start with simple things, such as the fact that no square number can end in 2, 3, 7, or 8, and that the last digits of squares, as you go up the ladder, endlessly repeat the palindromic cycle 1, 4, 9, 6, 5, 6, 9, 4, 1, with a 0 separating each cycle. [See Answers, Ten, opening section.] We show how to find the digital root of a number by adding its digits and casting out 9's until only one digit is left. All squares, they discover, have digital roots of 1, 4, 7, or 9. These also repeat a palindromic cycle, 1, 4, 9, 7, 7, 9, 4, 1, only now the cycles are punctuated by 9's instead of 0's."

"Those are useful rules for puzzle buffs," I said. "I remember one instance. I knew that 12,345,678,987,654,321 is the square of 111,111,111 and I wondered if 98,765,432,123,456,789 could also be a square. I wasted twenty minutes extracting the square root before I remembered the digital-root test. Since the number has a digital root of 8, it can't be a square."

Dr. Matrix nodded. "We also teach patients that if a square's last two digits are alike and not 00, they must be 44, and that 144 is the smallest example."

"Can a square end in three 4's?"

"Yes. The smallest is 1,444. Then comes a big jump. You might ask your readers to see if they can find the next smallest example—perhaps a formula that generates all such numbers." (See Answers, Ten, I.)

"Excellent," I said, jotting this in my notebook. "And do any squares end in four 4's?"

Dr. Matrix shook his head: "Three is the maximum. Since neither 44 nor 444 is a square, we see that no square of two or more digits can have all its digits alike. It is also true that no square of two or more digits can have all its digits odd. Your readers might enjoy proving it." (See Answers, Ten, II.)

When I had finished writing this down, Dr. Matrix said: "Here's a little-known curiosity about 13. Its square is 169. The reverse of 169 is 961, and the square root of 961 is 31, the reverse of 13. The product of 169 and 961 is 162,409, another square. The sum of the digits in 169 is 16. The sum of the digits in 13 is 4, the square root of 16."

"It's too much! You're blowing my mind!"

"There's a topper," Dr. Matrix said, smiling slightly. "If we ignore palindromic numbers such as 11 and 22, and numbers ending in 0 such as 10, 20, 30, then there's one other pair of two-digit numbers with the same set of properties as 13 and 31."

"I'll ask my readers that too," I said. (See Answers, Ten, III.) "I know a great deal of work has been done on square numbers that contain all the digits just once, with or without 0, and all of them twice, and so on. Have you done any work along such lines?"

"No, but let me mention a remarkable discovery sent to me recently by J. Malherbe, a Parisian mathematician I knew many years ago when I studied in France under the great Bourbaki. The two numbers 57,321 and 60,984 together contain the ten digits. And each of the squares of those numbers, 3,285,697,041 and 3,719,048,256, is made up of just the ten digits." (I later learned that there are three other pairs of numbers with the same property: 35,172 and 60,984; 58,413 and 96,702; and 59,403 and 76,182.)

"Can a square be exactly twice another square?"

"No. Nor can it be any prime multiple. But I'm more interested in less familiar corners of square theory. Have you heard of automorphic numbers?"

I shook my head.

"An automorphic number is one that reappears at the tail end of its square. For example, 5 and 6 are the only single-digit automorphs except for the trivial cases of 0 and 1. The two-digit automorphs are 25, with a square of 625, and 76, with a square of 5,776. The three-digit automorphs are 625 and 376. Our patients find these numbers symbolic of the fact that, even though they'll soon acquire the facade of a square, they can preserve their unique interior identity."

"I observe," I said, "that larger automorphs are obtained by adding digits to the front of the two automorphs of next lowest order. Are there two automorphs for any given number of digits?"

"Two at the most but sometimes only one. For example, 9,376 is the only four-digit example. We assume, of course, that no number begins with 0. And 90,625 is the only five-digit automorph."

"Are there automorphic numbers in all number systems?"

"No. If the base is a prime or a power of a prime, there are no automorphs except 0 and 1."

I thought for a moment. "Then base 6 would be the first system with true automorphs?"

"Yes. Base 10 is next. Maybe your readers would like to search for the two two-digit automorphs in base 6." (See Answers, Ten, IV.)

"Is there a limit to an automorph's size?"

Dr. Matrix assured me there was not. He borrowed my notebook and wrote from memory the hundred-digit base-10 automorph shown in Figure 13. There is a beautiful relation, he showed me, between pairs of automorphs of

3	9	5	3	0	0	7	3	1	9
1	0	8	1	6	9	8	0	2	9
3	8	5	0	9	8	9	0	0	6
2	1	6	6	5	0	9	5	8	0
8	6	3	8	1	1	0	0	0	5
5	7	4	2	3	4	2	3	2	3
0	8	9	6	1	0	9	0	0	4
1	0	6	6	1	9	9	7	7	3
9	2	2	5	6	2	5	9	9	1
8	2	1	2	8	9	0	6	2	5

Figure 13. One hundred digits of the automorphic sequence
ending in 5

the same length. If you know one, you can immediately write the other. Readers are invited to search for this relation and use it to construct the second hundred-digit base-10 automorph. (See Answers, Ten, V.)

"Of course most of our hippies, particularly the chicks, get a headache when they study arithmetic," Dr. Matrix said. "You'd be surprised, though, how many of them develop a strong interest in squares once they get over their hippie withdrawal symptoms." He glanced at his square wristwatch. "Holy McNamara! Time for lunch. You'll join us?"

Dr. Matrix, Iva, and I walked across the large quadrangle framed by the forty-nine houses of Squaresville. It was mid-November but the grass was still green and impeccably cut. Square signs read "Keep off grass," but Iva explained that this referred to marijuana and did not prohibit our walking diagonally across the lawn.

When we entered the dining room of Eisenhower Hall, 361 ex-hippies were standing and singing "America the Beautiful" against a backdrop of an enormous American flag that covered an entire wall. Iva led us to a square table in a section reserved for the staff. Dr. Matrix gave a blessing and everyone sat.

"What's their program for the rest of the day?" I asked.

"Today's Saturday," Iva replied. "There are no classes. The boys and girls will attend a showing of *The Sound of Music*, then go back to their rooms to watch television, and read *Reader's Digest*. There's a square dance this evening with music by Lawrence Welk and his orchestra. Tomorrow morning everyone goes to chapel. The sermon is usually preached by our staff minister, but tomorrow we have a guest minister. Dr. Norman Vincent Peale is preaching on 'Get in Gear with God.'"

"Do you have any trouble finding prominent guest speakers?"

"No, they're always pleased to help out in a square cause. Last week we had a sermon on the evils of Methedrine by Dr. Ellis D.—the hippies call him 'L. S. D.'—Sox, the public health director of San Francisco. Next week Dean Rusk will tell us why we are in Vietnam."

After coffee and a square-cut slice of apple pie the boys and girls stood up to chant the sixteen-word Squaresville law: "Good squares are trustworthy, loyal, helpful, friendly, courteous, kind, obedient, cheerful, thrifty, brave, clean, and reverent." After a chorus of "God Bless America" Dr. Matrix dismissed them.

The cats and birds looked clean and reverent as they filed politely past us. Each wore a square button. I noted one of the legends: "Love not Haight."

Dr. Matrix pointed to a clean-cut young man who saluted us as he walked by. "You should have seen him when he first arrived. He was a Digger in the East Village who called himself Launcelot—barefoot, with long pigtails, steel-rimmed glasses, and L-U-V painted on his forehead with lipstick. For three weeks all he ever said was 'Hey man.' His father in Tulsa had been sending him bread but he flung it all in $10 bills at the heads of brokers on the floor of the Stock Exchange. When he wired home, 'No mon, no fun, your son,' his father wired back, 'How sad, too bad, your dad,' and asked the New York fuzz to pick him up and bring him here. He was a typical meth monster until we switched him from meth to math. Next week he leaves to take a job as an IBM computer programmer at their research center west of here in Yorktown Heights."

I was much impressed and promised Dr. Hawk and

Mary Jane I wouldn't tip their identities until I got the word. Permission came near the end of December. The hippie movement was by then in such a shambles that the plastics, the part-time hippies from wealthy middle-class families, were no longer grooving. The hippies still hung up on the scene had parents who couldn't afford Squaresville's fee. Dr. Hawk sold the entire property to *Reader's Digest*. The forty-nine split-level homes are now filled with happy folk who commute daily between Squaresville and Pleasantville.

11. Left Versus Right

In November 1967, a year before the presidential election, Dr. Matrix was interviewed in New York City by Charlie Rice. The interview appeared in Rice's weekly column, "Punchbowl," in the Sunday newspaper supplement *This Week*. A supporter of Nelson Rockefeller followed up Rice's account with a second interview. It was the basis for a press release that was picked up by the wire services and widely reprinted in newspapers around the country. The original release read as follows:

"Numerological analysis of the last names of the leading presidential hopefuls for 1968 discloses that Nelson Rockefeller is the best 'balanced' politician in the United States, and Lyndon Johnson the most 'unbalanced.' This was disclosed today by Dr. Irving Joshua Matrix, the world's leading numerologist.

"To determine a political leader's left-right 'balance,' Dr. Matrix explained, one assigns to each letter of his last name the number giving the position of that letter in the alphabet. Thus *a* is 1, *b* is 2, *c* is 3, and so on. The sum of the letters on the left side of the name is then compared with the sum of the letters on the right to obtain the name's 'balance ratio.' For example, *Rom*, the first three letters of Romney, are 18, 15, and 13, which have a sum of

46. The last three letters, *ney,* are 14, 5, and 25, which add to 44. The name is balanced 46/44. This ratio indicates, Dr. Matrix said, that George Romney is fairly well balanced but leans slightly to the left.

"If a name has an odd number of letters, as in the case of Nixon, the central letter, called the 'fulcrum,' is ignored because its weight contributes to neither side. *Ni* has a sum of 23, *on* a sum of 29, indicating that Nixon leans 6 points to the right. William Buckley, though not a presidential candidate, has a 26/42 ratio, a large unbalance of 16 points to the right. This ratio is exceeded however, Dr. Matrix pointed out, by Dr. Benjamin Spock's unbalance of 35/14, or 21 points to the left.

"Among leading contenders for the presidency, Dr. Matrix declared, President Johnson is the most unbalanced. *Joh* adds to 33, *son* to 48, a strong bias of 15 points to the right. The only completely balanced contender is Nelson Rockefeller, who has a ratio of 52/52. '52 is an unusually heavy weight,' Dr. Matrix said, 'indicating that Rocky would draw heavy support from both right and left voters.'

"When asked about Shirley Temple, Dr. Matrix smiled and pointed out that 'Temple' works out to 38/33, a slight leftward list for the Good Ship Lollipop. But, he added, Shirley's campaign for Congress was under her married name of Black. This indicates, he said, that among all the dark-horse possibilities for president, Shirley is obviously the darkest. 'The name Black does balance with a 14/14 ratio,' he admitted, 'but the weight on both sides is so small that I'm afraid Mrs. Black must be considered a lightweight contender.'" [Mrs. Black is now, appropriately, the U.S. ambassador to the black nation, Ghana.— M.G.]

As it turned out, Nixon became the Republican presidential candidate, with Spiro Agnew (20 points right!) as his running mate. Opposing him was Hubert Humphrey (a mere 2 points left), with Edmund Muskie (28 points left!) as running mate. No wonder that the voters, who were in a strongly conservative mood, elected Nixon and Agnew.

12. Fifth Avenue

After Dr. Matrix sold Squaresville I lost track of his whereabouts until November 4, 1968. It was the day before Election Day and Iva telephoned to say that she and her father were in New York for a few days. Could I join them the next day for dinner? Of course I was delighted. We arranged to meet at three o'clock on the Promenade at Rockefeller Center, the little street that runs from Fifth Avenue to the gilt statue of Prometheus in the fountain by the lower plaza.

It was a light gray, overcast but pleasantly mild afternoon when I found the two of them sauntering counterclockwise around the breezy, beflowered Promenade. Dr. Matrix was his usual imposing figure—tall, gray-haired, his alert green eyes observing everything with intense interest. Iva was wearing her black hair upswept and lacquered; her flapping miniskirt caught the eye of every male on the Promenade. An exotic perfume diffused up my nasal passages as we exchanged affectionate kisses on the cheek.

We found an empty wooden bench in front of the French bookstore on the Promenade's downtown side. They had just returned, Iva told me, from Djakarta in Indonesia, where her father had been invited by the governor to advise him on the operation of Hwa Hwee, the city's popular numbers game. I recalled reading in the *New York Times* (Sunday, June 9, 1968) that since this ancient Chinese gambling game had been legalized in Djakarta early in 1968, the city's 4 million inhabitants had become so obsessed by it that even the country's appalling economic and political troubles had been forgotten.

Without going into the elaborate ritualistic details, Hwa Hwee begins each day at 11:00 A.M., when a number from 1 through 36 is selected in great secrecy. The number is placed in a cylinder; the cylinder goes into a red cloth bag; the bag is hung from the roof of a Buddhist "praying room" inside a small gambling hall in Glodok, the Chinese district. At intervals throughout the day a series of cryptic clues are issued to the public. Betting continues until 11:00 P.M. Precisely at midnight the winning number is announced by messengers who ride scooters through the city calling "Hwa Hwee!" followed by the winning number. Minimum bets are 250 rupiahs (about 75 cents), and all winners are paid 25 to 1. Profits to the city, which are enormous, have been used mainly to build new schools. For three months Dr. Matrix had been working with Tan Eng Giap, who provides the riddles for each day's number, on ways to guard against cheating and to discourage the rise of "black Hwa Hwee," illegal betting spots that were springing up everywhere in poorer sections of the sprawling city.

"Mr. Giap, whose initials backward spell GET, is a clever man," said Dr. Matrix. "I met him years ago when he was a croupier at an illegal gambling house in Djakarta. His clues are quite ingenious."

"I wish I could stay longer," Iva said to me as she stood up, "but there's some shopping I must do. We'll be meeting again in a few hours. Remind me to tell you about the trouble we once got into in Chicago when Father was hired by the Syndicate to give them advice about *their* numbers game."

"I presume," Dr. Matrix said after Iva had left, "you'll want to ask me about the outcome of today's voting."

"You presume correctly."

"Nixon undoubtedly will be the 37th president. Note that 37 is a prime number.* Interesting, because the only other president who was a Quaker was Herbert Hoover, and he was the last prime-number president, the 31st."

I began to write in my notebook while Dr. Matrix continued his analysis. "Nixon's great advantage, of course, is that his name ends in *on*. There've been nine previous *on* presidents, from George Washington to Lyndon Johnson, but only one *ey* president, William McKinley. Nixon's been counting heavily on this. That's why he kept stressing the slogan 'Nixon's the *one*.' Wallace will do fairly well because he alone has the double letter in his name, like so many earlier presidents of this century. Humphrey's initials, *HHH*, with their upside-down and mirror-reflecting symmetry, will get him more votes than he would otherwise have got and that should make it a close decision. In addition, *H* is the eighth letter of the alphabet and 888 also is the same upside down and mirror-reflected. The three 8's add up to 24, the sum of the digits in 1968. Unfortunately, all of this is insufficient to counter the stronger advantages of *on* over *ey*."

"Don't you think it strange," I said, "that the last two significant third-party movements have both been led by a Wallace? Henry Wallace, left of center, in 1948. Now George Wallace, right of center, just twenty years later."

---

* For the numerology of 37 see Charles W. Trigg, "A Close Look at 37," *Journal of Recreational Mathematics* 2 (1969):117–128.

"Paralleled, of course," said Dr. Matrix, "by the equally strange right-left reversal involving two other last names, Joseph McCarthy and Eugene McCarthy. Have you noticed that the first names of the two Wallaces combine to make Henry George? He ran for mayor of this city in 1886—note that 86 is 68 backward—on his single-tax program. It was as simplistic and self-defeating as the two Wallace movements."

Dr. Matrix had other curious comments on the names of the three presidential candidates but space allows me to give only the following word puzzles, which he said he had prepared specially for my use. They belong to a class he called "minimal king's-tour spelling matrices." Old puzzle books often contain rectangular matrices with a letter in each cell. By moving from cell to cell in the manner of a chess king one tries to spell out a proverb, or as many names as possible of flowers, animals, and so on. Most puzzles of this type are on the dull side, but Dr. Matrix had thought of a way to make them combinatorially less trivial. The idea is to take the full name of a prominent person, a name preferably with no double letters (although double letters can be accommodated by the rule that a cell may be counted twice), and design a symmetrical pattern of square cells on which the name can be spelled by a single "king's tour." There is one further proviso: the pattern must be minimal in the sense that no letter appears on it more than once.

Dr. Matrix then showed me two remarkable examples (see Figure 14). The line on the pattern at the top indicates how a king can be placed on L and then moved one square at a time in any direction (as the king moves in chess) to spell LYNDON BAINES JOHNSON. On the lower left HUBERT HORATIO HUMPHREY can be spelled in the same way. Can the reader place the thirteen different let-

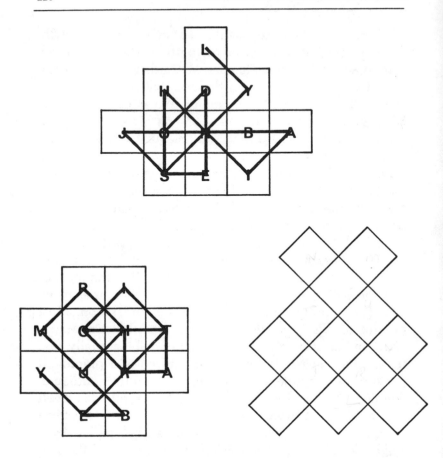

Figure 14.   Three spelling matrices. Fill the one at bottom right
to spell RICHARD MILHOUS NIXON

ters in RICHARD MILHOUS NIXON on the cells of the pattern
at bottom right so that a king's tour will spell the new
president's full name? (See Answers, Twelve, I.)

   "I assume you have read Edward Jay Epstein's eye-opening hatchet job on Big Jim Garrison, the New Orleans district attorney, in the *New Yorker* [July 13, 1968]," * I said. "It explained how Garrison derived Jack Ruby's unlisted Dallas telephone number from that mysterious number 19106 in Lee Harvey Oswald's notebook." (The digits in 19106 are rearranged by taking the first digit, then the last, then the second, then the second from last, and finally the center digit. This produces 16901. Subtracting 1300 yields 15601, which was Ruby's private telephone number.)

Dr. Matrix's emerald eyes glittered with amusement. "Extremely crude numerology. A much simpler way to decode 19106 is to partition it 1–9–10–6. Using the familiar code of *A* equals 0, *B* equals 1, *C* equals 2, and so on, the four numbers translate as *BJKG*. The *K* obviously stands for Kennedy and the remaining letters, *BJG*, are the initials for Big Jim Garrison. If you prefer to partition the number 1–9–1–0–6 you get *BJBAG*, or Big Jim's Bag." The same code, Dr. Matrix later informed me, also applies to 18960, the difference between Oswald's 19106 and 00416, the prison number of James Earl Ray before he escaped from Missouri State Penitentiary in 1967. 1–8–6–9–0 decodes as *BIGJA*, the beginning of "Big James."

"I must write Mort Sahl about this," I said, chuckling as I recorded the details.†

We stood up and began a leisurely stroll along Fifth Avenue.

"The coming year should be interesting," Dr. Matrix

---

* The basis of Epstein's book *Counterplot*, published by Viking in 1969.

† At the time, stand-up comic Sahl was vigorously standing up comically in support of Garrison's absurd conspiracy theory.

remarked; "69 is the same upside down. An appropriate number, don't you think, for the ending of the 'sexties'—or should we call them the 'sicksties'? 69 backward is 96 and SIXTY-NINE and NINETY-SIX are anagrams."

We paused to look into a window in which copies of James D. Watson's recent best seller *The Double Helix* were displayed. "Crick, Watson, *and* Wilkins shared their 1962 Nobel prize," Dr. Matrix mused. "Have you noticed that AND is DNA backward?"

I pointed up the street to the enormous numerals 666 on top of the Tishman Building at 666 Fifth Avenue. "Any comments?"

"We've discussed the Book of Revelation's Mark of the Beast before," he said with a sigh. "Frankly, the topic bores me. A skillful numerologist such as myself can easily find 666 in anybody's name. Consider Vincent Lopez, the orchestra leader who fancies himself a numerologist. [Lopez's book *Numerology: How to Be Your Own Numerologist* was published by Citadel Press in 1961.] I once amused myself by numbering the alphabet backward, starting with 101 for Z and ending with 126 for A. In this code the letters of V. LOPEZ have the values 105, 115, 112, 111, 122, 101. They add to 666. Or take my own full name, Irving Joshua Matrix. Each name has six letters. And you'll find my middle name, as well as a clue to who I really am, in the Bible's sixth book, sixth chapter, sixth verse. Amusing, but of course trivial."

"You once told me," I said as we waited for the traffic light at 52d Street to change, "that every integer has unique numerological properties."

"Naturally."

I pointed to the street sign. "How about 52?"

"It's the number of cards in a deck," Dr. Matrix answered promptly, "as well as the number of weeks in a year. You might be surprised to know that if you take the names of the thirteen card values from ACE to KING and count all the letters you'll find there are exactly 52. The four suits correspond to the four seasons and the twelve picture cards to the twelve months. Add the values of all 52 cards, plus 1 for the joker, and you get 365, the days of the year." *

I aimed my mechanical pencil at a sign that said "Store open weekdays from 9 to 5" and asked Dr. Matrix what he could do with that.

"9 over 5, or ⁹⁄₅, added to the square root of ⁹⁄₅ is 3.1416+; a remarkable approximation to pi, is it not? That was discovered last year by my friend Fitch Cheney. You might ask your readers to write the first three digits of pi, 314, drawing the 4 so that its two nonhorizontal bars meet at the top, and then hold it up to a mirror. They'll get a pleasant surprise."

As we approached 57th Street we crossed the avenue and walked into Tiffany's. "Iva's going to meet us here in about ten minutes," Dr. Matrix said, glancing at his wristwatch. "I'm buying her an unusual bracelet for her birthday next month."

The bracelet Dr. Matrix had asked Tiffany's to make consisted of sixteen spherical beads, all the same size, half of them jade and half pearl. Instead of alternating pearl and the jade, however, the beads were strung in what appeared to be a haphazard manner (see Figure 15).

"Do you see the order behind this arrangement?" Dr. Matrix asked.

I studied the bracelet for several minutes before admitting that I could perceive no order at all.

---

* The number 52 plays an important numerological role in Vladimir Nabokov's *Lolita*. See notes 253/14 and 253/15 in *The Annotated Lolita*, edited by Alfred Appel, Jr. (New York: McGraw-Hill, 1970).

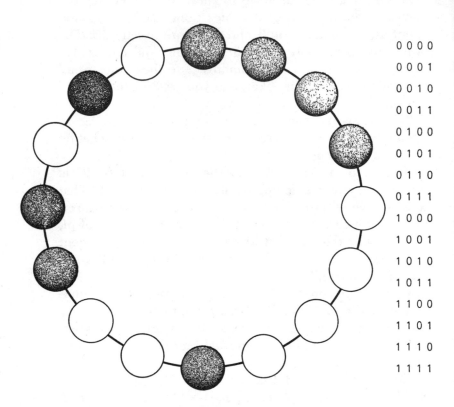

Figure 15. Combinatorial bracelet with sixteen quadruplets of
jade (1) and pearl (0) beads

When a penny is flipped *n* times, Dr. Matrix explained,
the number of equally probable ways it can fall is 2 to the
power of *n*. Two flips have four possible outcomes: HH,
HT, TH, and TT. Three flips have eight possible outcomes, four flips have sixteen, and so on. The same applies
to the different arrangements of *n* beads in a row when
each bead can be either of two different colors. An interesting combinatorial problem now presents itself. Is it pos-

sible to arrange 2^n beads in a circle, half of them one color
and half another, so that every possible n-tuple combination will be represented by n adjacent beads as you go
around the circle?

The answer is yes. There is a unique solution when n
equals 2 (see Figure 16). If you go around the circle in either direction, taking the beads two at a time, you find all
four possible doublets. When n equals 3, the solution is
also unique. Circle either way, taking adjacent beads three
at a time, and you encounter all eight possible triplets. Of
course, the beads can be arranged in reverse order, but
since a reversal is obtained merely by turning over the
bracelet or circling it in the other direction it is not considered different. You might suppose a new triplet solution
could be obtained by taking the "complement" of this solution—changing each white bead to colored and each colored bead to white—but actually the resulting bracelet is
identical with the one shown.

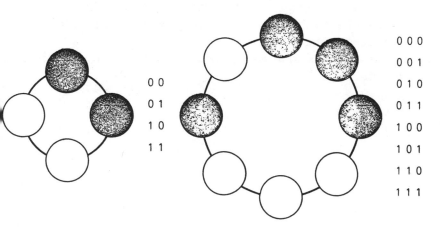

Figure 16.   Combinatorial bracelets for doublets (left) and triplets (right)

The minimal-length bracelet showing all possible quad-ruplets of jade and pearl beads will have 2^4, or sixteen, beads. It was such a bracelet that Dr. Matrix had prepared for Iva. Go in either direction around that bracelet (Figure 15), taking the beads four at a time, and you will be able to check off all sixteen combinations of 1 and 0 (with the 1 standing for jade and the 0 for pearl, or vice versa); the six-teen arrangements correspond to binary forms of the numbers 0 through 15.

Reversing the order of the beads provides no new solu-tion. In this case, however, the complement (obtained by interchanging colors) does produce a new solution. There are eight basically different arrangements that solve the triplet problem, considering complements as being dif-ferent but not reversals. The reader knows of two: the one shown and its complement. Can he find the other six? (See Answers, Twelve, II.)

On the busy sidewalk outside Tiffany's, while Iva was admiring her bracelet, Dr. Matrix extended his hand. "Sorry I have other plans and can't join you two for dinner and the evening."

"So am I," I said as we gripped hands, but I didn't really mean it.

13. The Moon

After the historic completion in July 1969 of the *Apollo 11* moon mission it occurred to me that Dr. Matrix might have some enlightening comments about it as well as some insights into the approaching end of the 1960–69 decade. I sent a letter to the last address I had for him, but the peripatetic doctor and his daughter Iva had already left for parts unknown. My letter was forwarded five times, to various European and Asian cities, before it finally caught up with him.

I had asked Dr. Matrix a series of questions. They follow, together with slightly cut but considerably edited versions of the great numerologist's replies.

QUESTION: Have you observed any remarkable numerological aspects of the *Apollo 11* mission?

DR. MATRIX: The key symbol of the "moon landing" (note the 11 letters) is, of course, 11. It can be interpreted in two ways: as the number 11 and as a pair of 1's.

Consider it first as 11. The 11th letter of the alphabet is *k*. Surely it is no coincidence that President Kennedy initiated the Apollo project and that the lift-off of *Apollo 11* was from Cape Kennedy. 11 is the smallest prime factor of 1,969. The landing was in the Sea of Tranquility and "tran-

quility" * has 11 letters. The first message received from
lunar soil, Neil Armstrong's "That's one small step for
man, one giant leap for mankind," has exactly 11 words.
(Armstrong later said he'd said "for A man," but so strong
is 11's power that the message was received without the
A.) Armstrong was then 38 and 3 plus 8 is 11. When the
three lunar explorers splashed down in the Pacific, they
landed 11 miles from the recovery ship, the carrier *Hornet*.
On the carrier they were given buttons to wear that read
"Hornet plus three." The six letters of HORNET added to
the five letters of THREE equal 11. These are only a small
fraction of the 11's involved in the *Apollo 11* flight.

Now consider 11 as a pair of 1's, symbolizing the first
two men to walk on the moon. Now, *a* is the first letter of
the alphabet. Is it not significant that the first two men to
put their footprints on the moon, Armstrong and Edwin
Aldrin, both have last names that start with *A?* Note also
the two *A*'s in NASA. If we write Armstrong's name as
Neil Arm Strong, Astronaut, we obtain NASA as an acrostic.

Andrei Voznesensky's clever Russian palindrome [re-
ported in *The New York Times*, July 21, 1969], A LUNA
KANULA, meaning "The moon has disappeared," was written, the Russian poet explained, so that one can travel letter by letter to the moon and back. Note that the palindrome has exactly 11 letters, which include four *a*'s and
the two *l*'s of APOLLO. Armstrong's middle initial is *A*.,
so that the first men on the moon actually had three *A*. initials
between them; the palindrome's extra *A* is for Andrei.
There are also three *a*'s in "United States of America."
Perhaps it was because of the double *A*.'s in Armstrong's

---

* The spelling preferred by NASA and the *New York Times*.

initials, symbolized by 11, that he was the first to set foot on the lunar surface.

Edwin Aldrin (11 letters) has *E.* as a middle initial, providing two *E.* initials. We all saw the astonishing "ease" with which Aldrin bounded over the lunar soil. His youngest child, Andrew, is 11 and has the initial *A.* for his first name as well as his last. Edwin Aldrin's mother's maiden name was Moon. Colonel Michael Collins (Mike Collins has 11 letters) was the astronaut who remained to pilot the command module. Can it be coincidence that his last name, COMMAND, and COLUMBIA, the name of his ship, all begin with *c?*

As I write, *Apollo 12* is planned for November, with the moon walk to be made by astronauts Alan L. Bean and Charles Conrad. Thus the *a* walk of *Apollo 11* will be followed by the *b* and *c* walks of *Apollo 12*, the *b* being symbolized by the 2 of 12, the *c* by the sum of 1 and 2. The *Apollo 12* mission should be as easy as *ABC*, but *Apollo 13*, with its unlucky number, is fraught with danger.*

Finally, your readers may enjoy finding the appropriate single 11-letter word that can be formed by rearranging the 11 letters of MOON STARERS. [See Answers, Thirteen, I.]

QUESTION: Do you have a digital problem related to space travel?

DR. MATRIX: No sooner had the three astronauts returned to the earth than Vice-President Agnew proposed

---

* The numbers 11, 12, and 13 all played crucial roles, Dr. Matrix later informed me, in the ill-fated Apollo 13 mission. It was launched on April *11*, 1970. Each of the three astronauts—James A. Lovell, Fred W. Haise, Jr., and John L. Swigart—had a name of *12* letters. But on April *13* the mission was aborted because of an oxygen tank explosion, and the crew returned to earth. JAMES, FRED, and JOHN have 13 letters among them.

that the U.S. begin immediate plans for a landing on Mars before 2000. It is numerologically significant that among the 10 letters of SPIRO AGNEW no letter is repeated. Multiply SPIRO by the mystical digit 7—the seven-day week of creation mentioned by President Nixon in his ebullient welcome to the returning astronauts in July, the seventh month—and let AGNEW be the product. (The total number of symbols is now 11.) This produces the following cryptarithm:

$$\frac{\begin{array}{r} \text{S P I R O} \\ 7 \end{array}}{\text{A G N E W}}$$

Assuming that each letter stands for a different digit and, as is customary in such problems, that neither five-digit number begins with 0, the cryptarithm has one and only one solution. Your readers may derive pleasure from finding it. You can see at once that S must be 1; otherwise the product would have more than five digits. For the same reason P must be either 0, 2, 3, or 4. Further reflection will show that A must be 7, 8, or 9, and that O cannot be 0, 1, 3, or 5 (1 would duplicate the 1 of S, 3 would cause W to be 1, 0 or 5 would duplicate O and W). The problem becomes tougher from here on, although it can be solved easily without the help of a computer. If the multiplier is 0, 1, or 6, there is no solution. Any other single-digit multiplier except 7 has more than one solution. [See Answers, Thirteen, II.]

QUESTION: The end of 1969 will terminate the 1960s. Do you have any numerological comments?

DR. MATRIX: As you may recall, you once quoted me as saying that the decimal expansion of pi, properly interpreted, conveys the entire history of the human race. The

1960s have been ten years of unprecedented change and variety. The ten decimals of pi from the 60th through the 69th are 4592307816.

What is significant about 4592307816? They are the first ten consecutive digits in pi that contain each digit exactly once! It is infuriating that the first two digits are not reversed; then there would be alternating odd and even digits for the entire series. The 70th decimal is 4, the same digit that begins the series. I predict that a major event, involving the number 4 and closely related to a 4-event of 1960, will occur in 1970.*

Computer expert Donald E. Knuth has called my attention to another fantastic fact that strengthens my interpretation of pi's decimals 60–69. The 1,960th through the 1,969th decimals are 5739624138. Here there is no 0 because of the repetition of 3, but all digits from 1 through 9 appear. The chance of finding all ten digits in a given sequence of ten consecutive random digits is, Knuth tells me, $10!/10^{10}$, or .00036288 exactly. Naturally the probability of finding this as early in pi as decimals 60–69 is not that small, but it is still small enough to deserve earnest meditation.

The decimals of pi from the 70th through the 79th are almost as fascinating: 4062862089. Observe the curious repetition of the same four digits in adjacent middle quadruplets and note that the first nine digits are even.

---

* This was dramatically confirmed on the *fourth* day of May 1970, when *four* students at Kent State University in Ohio (note that both KENT and OHIO have *four* letters) died after being shot by National Guardsmen. It was the climax of student protests against the Vietnam War. 1960 was the year that students began their previous great wave of protests against racial discrimination in the South. The first such protest was a sit-in on February 1 when *four* black students in Greensboro, North Carolina, refused to move from a lunch counter where they were denied service.

Something exceedingly odd undoubtedly will take place in 1979.

One final observation on 1969. Square it, circle the last four digits of the product, then view them upside down. You will see 1969! Aside from the trivial case of the year 1, how many years from 1 through 9999 have the property of appearing upside down at the tail of their square? [See Answers, Thirteen, III.]

QUESTION: Can you provide a problem of nontrivial interest that ties in with moon-exploration plans?

DR. MATRIX: Assume the moon is a perfect sphere and that we want to establish n lunar bases as far apart from one another as possible. More precisely: How can n spots be arranged on a sphere so that the smallest distance between any pair of spots is maximized? The problem is equivalent to that of placing n equal, nonoverlapping circles on a sphere so that the radius of each circle is maximum.

When $n = 2$ the answer is obvious: the two bases are placed at opposite ends of a diameter. For three bases the solution is to put them on an equator at the corners of an inscribed equilateral triangle. For four bases the unique solution is the corner points of an inscribed tetrahedron.

Let us jump to $n = 6$. Here the only answer is to place the spots at the corners of an inscribed regular octahedron or, what is the same thing, at the centers of the six faces of a circumscribed cube. The case of $n = 5$ has an unexpected answer. The maximum distance turns out to be the same as when $n = 6$. Simply take the solution for six spots and remove one spot. The result is the best possible answer for five spots. The solution is not unique; two spots must be at opposite ends of a diameter but the other three can be shifted along the equator to an infinity of positions.

There are unique solutions for seven and nine spots.

When $n = 8$, the best placing of spots is not, as one might guess, at the corners of an inscribed cube but at the corners of a square antiprism. When $n = 12$, the corners of an inscribed icosahedron (or the centers of the faces of a circumscribed dodecahedron) give the best pattern. The solution for $n = 10$ was given by Ludwig Danzer in 1962, but the full proof remains unpublished. Danzer had earlier settled the case of $n = 11$ by showing it to be similar to $n = 5$. The best arrangement is obtained by removing one spot from the twelve-spot solution. The eleven spots [see Figure 17] are fixed. They cannot, as in the case of $n = 5$, be shifted about to provide other solutions.

The only other known solution is when $n = 24$. In 1961 Raphael M. Robinson proved a conjecture of B. L. van der Waerden's that the answer is the vertexes of an inscribed snub cube, one of the Archimedean semiregular solids [see Figure 18]. This polyhedron is asymmetric, so that it has left-handed and right-handed forms. Its thirty-eight faces consist of six squares and thirty-two equilateral triangles. [See Answers, Thirteen, IV, for further references.]

QUESTION: Anything else?

DR. MATRIX: November 1969 is what my friend Kirby A. Baker, a mathematician at the University of California in Los Angeles, calls a perverse month.

A perverse month is one that has six calendar weeks, requiring six lines on the calendar unless the last one or two days are shown by tiny numerals squeezed into the fifth line. What is the maximum number, Baker asks, of perverse months that can occur in a year? What is the minimum number? Both questions are closely related to the maximum and minimum numbers of Friday the 13ths that can occur in a year, which are three and one. As you may know, it has been proved that the 13th is more likely to fall on Friday than on any other day of the week. 1970

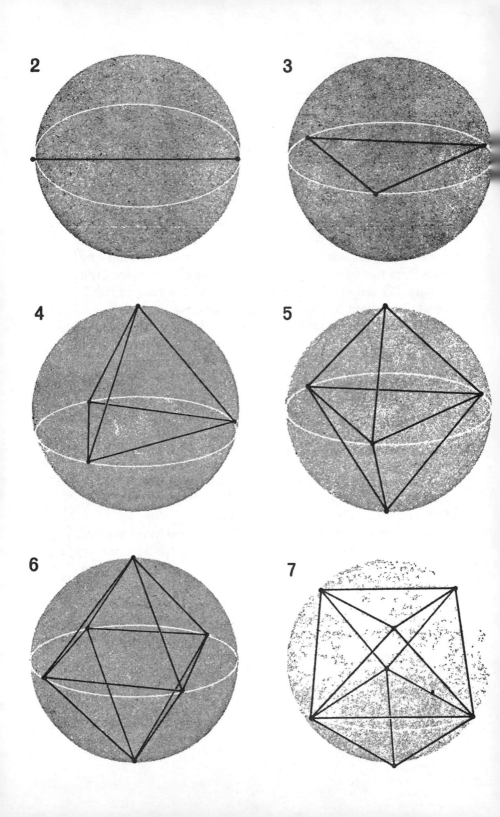

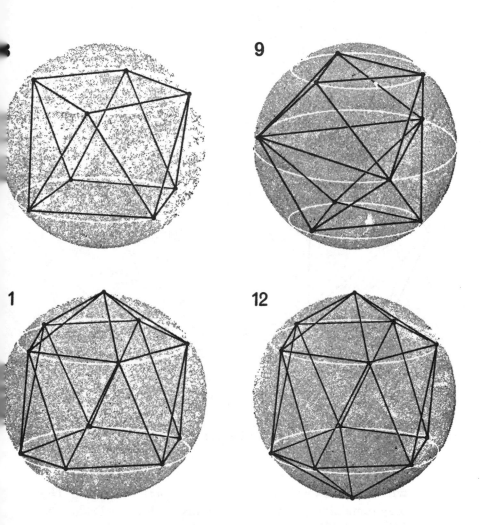

Figure 17. How to space lunar bases as far apart as possible. Solutions are given for two through nine, eleven, and twelve bases

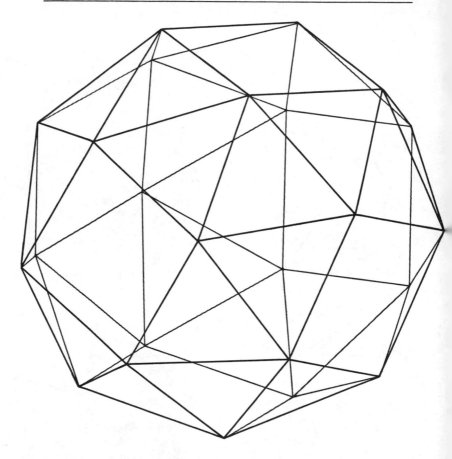

Figure 18. The snub cube, the twenty-four-base solution

will have the maximum number—an ominous omen. The only remaining years of this century with three Friday the 13ths are 1981, the notorious 1984, 1987, and 1998.

Baker has shown that the more perverse months, the fewer the Friday the 13ths, and the fewer the perverse months, the more Friday the 13ths. "Thus," he writes, "a year good for calendar makers is unlucky for the rest of us, and vice versa." [Can the reader determine the exact relationship between perverse months and Friday the 13ths in a given year? (See Answers, Thirteen, V.)]

14. Honolulu

In spite of the soft winds of Hawaii, its bright azure skies, and the friendly smiles of the islanders, I always have the feeling that strange mysteries lurk behind every window in this homeland of Charlie Chan. I was therefore not surprised, when I recently attended a mathematical congress at the University of Hawaii, to encounter Dr. Matrix as I was leaving the coffee shop of my Honolulu hotel. The eminent numerologist was finishing breakfast, his hawk-like nose buried in a Japanese newspaper. I did a double take and spun around to greet him.

"Yes, it's been more than two years since we last met," he said, standing up to shake hands. He was dressed nattily in a white linen suit and his gray hair was cut in fashionably long sideburns. Two viridescent eyes regarded me with quizzical amusement.

"Are you here for the mathematics gathering?" I asked.

"No. Iva and I are on our way to Tokyo, but we stopped off in Honolulu for three weeks of relaxation in the surf."

Iva was spending the morning on Waikiki Beach, just outside our hotel. Dr. Matrix said he expected her back before lunch. When he invited me to their rooms in the meanwhile, I accepted at once, even though it meant missing the morning's symposium on new developments in topology.

The first thing I noticed when we entered Dr. Matrix's spacious suite was a large chess table in the center of the room. A set of intricately carved black and white ivory pieces was set up for play.

"I didn't know you liked chess," I said.

"Yes. Unfortunately I play an excellent game. Wasn't it Montaigne who wrote that chess was too serious to be entertaining and too frivolous to be taken seriously?"

"I'm not sure, but I recall Philip Marlowe, in one of Raymond Chandler's detective novels—*The Long Goodbye*, I think—saying that chess is 'as elaborate a waste of human intelligence as you could find anywhere outside an advertising agency.'"

"Quite true," said Dr. Matrix. "Why don't we waste a little intelligence while we talk?"

We seated ourselves. Dr. Matrix picked up a white and a black pawn, rattled them in his cupped hands, then held out two fists. I tapped his right hand, but when he opened it, the hand was empty. Before I could say anything he closed his right hand and opened his left. It too was empty. He quickly closed his left hand, then opened both fists. A pawn rested on each palm!

"Just some sleight of hand I learned from Tenkai, the Japanese magician," Dr. Matrix explained, chuckling. "I worked for a while as his assistant when I was young. But you *did* pick the white pawn, so it's your move."

We carefully rotated the table to bring the white pieces around to my side. "I'm not a good player," I said, shoving my king's pawn forward two squares.

Dr. Matrix made the same move with his king's pawn. I jumped my knight to king's bishop three, then Dr. Matrix protected his attacked pawn in the usual way by moving his queen's knight to her bishop three. I placed my king's bishop on bishop four—the old Giuoco Piano opening.

"What's new in numerology?" I asked.

"The inevitable correlations," he replied. "Have you noticed how many coincidences have been turning up lately—of course they're not really coincidental—between the names of famous people and their activities?"

I shook my head as I took a pencil and small note pad from my shirt pocket.

"Well, take Lionel Tiger, for instance. He's the anthropologist who's becoming so well known for his writings on animal behavior and its relation to human behavior. His round face even *looks* a bit lionish or tigerish. Tiger's been working closely with his colleague at Rutgers, the anthropologist Robin Fox. They first met, you know, at the London zoo. Then there's Miss Iris C. ["I see"] Love, an archaeologist at Long Island University. In 1969 she unearthed the Temple of Aphrodite at Cnidus in Turkey. Last November she identified what may be the original head of the famous nude statue by Praxiteles—the temple's main attraction. I've always regretted that in my previous incarnation as Pythagoras I lived too early to see it."

"As I recall," I said, "Miss Love found the head in the dark, dusty depths of the British Museum."

"Yes, and did you notice that the museum had given it the catalog number 1,314? Of course 1,314 is a cyclic permutation of the first four digits of pi, the most elegant of all ratios. The circle has always been *the* symbol of female beauty. Last August a book called *Oh! Sex Education* caused quite a stir. It was written, naturally, by Mary Breasted."

"I've been reading a new book on the menace of noise pollution," I said. "It's by Henry Still."

"Yes," said Dr. Matrix, "a person's name exerts a strong influence over his interests. "There's J. J. C. Smart, the smartest philosopher in Australia, and Kuan-han Sun, a

Chinese physicist at Westinghouse who's doing research on the influence of the solar wind on the moon. Anton Horner was the solo horn player with the Philadelphia Orchestra for twenty-eight years. James Cash Penney founded the Penney chain stores and became a billionaire. Are you familiar with Dover's paperback reprint of *Solid Geometry* by L. Lines, or Cambridge University Press's edition of *The Theory of Ruled Surfaces* by W. L. Edge?"

Dr. Matrix gave me no time to answer. "I could go on and on," he said. "Have you ever considered the meanings in German of the names of the great pioneers of psychoanalysis? Freud(e) is the 'pleasure' of the 'pleasure principle.' Jung is the 'young' rival of Freud. Stekel means 'little stick,' an obvious sexual symbol. Adler is the eagle who soars above them all. Was there ever a more appropriate name in German for an analyst than Horney?"

Dr. Matrix paused until I had written all this down.

"A shoe company in St. Louis," he resumed, "is owned today by a man named Shoemaker. Dr. Shuffle is a podiatrist in Washington, D.C. There are two gynecologists in Boston who are partners: Dr. Hands and Dr. Fingerman. Several men named Glasser are in the glass and glazing business in New York and . . ."

"I was told recently by Leon Svirsky, the science writer," I interrupted, "that Philip Morrison, the physicist who edits the *Scientific American*'s book review section, is a friend of Morris Philipson, the director of the University of Chicago Press."

"But of course," Dr. Matrix said. "Only a dull-witted iconoclast would suppose such patterns are accidental. You find them everywhere, particularly in connection with creative people. Consider the traditional spelling rule, '*i* before *e* except after *c*, or when sounding as *a*, as in NEIGHBOR and WEIGH.' The rule is broken twice by AN-

CIENT SCIENCE. Then along came Albert Einstein, who broke so many basic laws of ancient science. His last name also violates the rule twice."

Dr. Matrix waited until I had written all this down before he continued.

"In July 1969 Dorothy Hodgkin, a British chemist, was the first to work out the exact three-dimensional structure of the insulin molecule. Observe the repetition here of the sacred digit 7. There are seven letters in INSULIN, DOROTHY, HODGKIN, and BRITAIN. July is the seventh month and 1969 has a digital root of 7."

I have space for only a portion of the rest of Dr. Matrix's analysis. He pointed out that insulin is a HORMONE, a PROTEIN that increases the rate at which GLUCOSE enters muscle tissue. The three words each have seven letters. "And the insulin molecule," Dr. Matrix said, "has precisely 777 atoms. But we're neglecting our game."

Dr. Matrix's third move caught me off balance. He placed his knight on his queen five square, leaving his king's pawn unprotected. Was it a deliberate gambit or a careless oversight? I could see no serious threat from his knight, and so I took his pawn with my knight. My bishop and knight were now both attacking his vulnerable king's bishop pawn.

Dr. Matrix ignored the attack. When he moved his queen out to knight four, I lost no time grabbing his bishop's pawn with my knight. The knight now threatened both his queen and his rook, so that I was certain of winning the rook. And Dr. Matrix had boasted of his chess prowess! I felt a twinge of embarrassment for him.

Dr. Matrix seemed unconcerned. "Some letter patterns are difficult to interpret," he said. "Consider the initials of Martin Luther King, Jr. They are *M.L.K.J.*—four consecutive letters of the alphabet in reverse order. That's most

unusual, but I'm not sure what it signifies. The terminal
letters of Martin Luther King, *nrg*, are not only the three
consonants of NEGRO, but when you pronounce them they
give the word ENERGY, one of Dr. King's outstanding
traits. Insert two vowels between *M.L.K.* and you get MA-
LIK. That's how the word for KING in both Arabic and Hebrew is pronounced."

"What about King's murderer, James Earl Ray?"

"The initial letters reversed, *R.E.J.*, suggest regicide,"
said Dr. Matrix, "the killing of a king. The terminal letters,
sly, have an obvious meaning, and if we include the last
two letters of his last name we get SLAY."

First and last letters of familiar sets of words, Dr. Matrix
explained, often conceal strange correlations. NEWS, which
comes from all directions, is spelled with the initials of
north, east, west, and south. The initials of the nine planets, in order from the sun, contain SUN in order. The last
letters of the names of the numbers 1 through 10 end with
TEN. Dr. Matrix mentioned a friend in Berea, Ohio, Theodore Katsanis, whose son Jason was born in September.
September is in the middle of five consecutive months
with initials that spell JASON. Edwin M. McMillan, a physicist, recently told Dr. Matrix about a California disk
jockey who worked on both FM and AM stations. His business card reads: "J. Jason, D.J., FM-AM." The twelve letters are a cyclic permutation of the initials of all twelve
months!

Dr. Matrix paused in his discourse to take my knight's
pawn with his queen. I had anticipated that. To save my
rook I was obliged to move it to the bishop's square.

"Mary McCarthy is a brilliant woman," Dr. Matrix went
on. "In her book *Cast a Cold Eye* there is a story called
'C.Y.E.' It tells about the inscrutable nickname, Cye, given
her by a group of classmates at a convent school. They

never told her what it meant and Mary tries to guess. The initials of Catch Your Elbow? Clean Your Ears? Clever Young Egg? It's curious Mary didn't see that her 'cold eye' is the clue. Take out *olde* and her youthful nickname is left."

Dr. Matrix took my king's pawn with his queen and said, "Check."

I gave a start. It was a ploy I had failed to see. The only way I could escape checkmate and not lose my queen was, of course, to interpose my bishop. I moved it back to the square in front of my king. The threatened mate was blocked, but now I would surely lose my king's knight. On the other hand, the black king would be exposed, and with an open file for my rook perhaps I could mount a quick counterattack.

"Can we move on to numbers?" I asked. "As you know, I like to give my readers little problems. Any curious number puzzles that are new?"

"In this age of digital computers," Dr. Matrix replied, "they turn up every month by the hundreds. Robert E. Smith, a mathematician with the Control Data Corporation, told me how a student at one of his company's institutes improved on an ancient result given by Plato. In his *Laws*, book 5, Plato recommends that a city be divided into plots of land so that the number of plots has as many proper divisors as possible. He suggests 5,040 because it has fifty-nine divisors, a rather large number. [This includes 1 but not 5,040.] Your readers who have access to computers, even some who don't, might like to know that the largest number of proper divisors a number less than 10,000 can have is sixty-three, which tops Plato's number by four divisors. There are just two such sixty-three-divisor numbers. One is 9,240."

"Delightful," I said. "I'll ask our readers to try to find the other one." (See Answers, Fourteen, I.)

"What's your room number?"

I took out my hotel key to check. It was a number with three digits.

"Remarkable," said Dr. Matrix, his green eyes squinting through a window and over the waving palm trees. "I trust you recognized it as a square number. Now, if you write it down, then put a certain other three-digit square number directly under it, you produce an unusual matrix. Each of the three two-digit numbers that make the three columns, when read from the top down, is also a nonzero square number."

After Dr. Matrix showed me the only solution (see Answers, Fourteen, II) he pushed a concealed button on the side of the chess table. The room was instantly filled with music.

"It's the score of Sir Arthur Bliss's 1937 ballet *Checkmate*," Dr. Matrix said as he lifted his knight and slammed it down on his king's bishop six square.

I stared at the board, my jaw sagging. It was indeed checkmate. My bishop, pinned by the black queen, could not take the checking knight, and there was no empty square next to my king. It was one of those rare events in chess, one that only knights can execute: a smothered mate (see Figure 19).

"Sorry about that, Gardner," said Dr. Matrix, "but I wanted to end the game with my seventh move. It's an old chess hustler's trap. I'm surprised you fell for it."

The hall door opened and Iva entered on bare feet, enveloped by a pale purple beach robe and a cloud of well-remembered perfume. "Aloha!" she called out, her black eyes flashing a smile. "How on earth did you know my father and I were here?"

"I didn't. My being in Hawaii now, believe me, is purely occidental. I hope you're free this evening to pilot an ignorant mainlander around the city."

Figure 19.   Dr. Matrix's smothered mate

"One more stupid pun," Iva said, wagging a scarlet-nailed forefinger in my face, "and the deal's off."

I was about to add that I had great difficulty getting myself properly oriented in Honolulu, and that I was more interested in bottomology than topology, but I kept my mouth shut.

15. Houston

> O speculators about perpetual motion,
> how many vain chimeras have you
> created in the like quest? Go and take
> your place with the seekers of gold!
>
> —Leonardo da Vinci

A startling letter from Bing Soph, an old friend now living in Houston, arrived in my mail early in August 1971. Soph informed me that an elderly man from Wales, calling himself Llewelyn Hooker, Jr., had created a considerable stir in Houston with his claim to have invented a perpetual motion machine. A small working model of the device was on display in the lobby of the plush Shamrock-Hilton Hotel, where Hooker and his pretty assistant, Miss Jacquelyn Jones, were selling stock for the Hooker Dynamaforce Corporation.

Soph enclosed the company's handsomely printed prospectus. It outlined Hooker's scheme to build a gigantic dynamaforce generator on the Houston Ship Channel. Not only would dynamaforce provide cheap, unlimited power for the area; it would cut the nation's air pollution in half as soon as the clean dynamaforce motor replaced the dirty internal-combustion engine. The prospectus contained no pictures of Hooker or Jones, but Soph had mentioned that the tall, bearded Hooker had green eyes and a prominent nose, and that Miss Jones's attractive features were unmistakably Eurasian.

I more than suspected that Hooker and Jones were none other than the notorious numerologist Dr. Irving Joshua

Matrix and his half-Japanese daughter Iva, but I spent the next hour shuffling around the letters of LLEWELYN HOOKER, JR., to see what else they might spell. Dr. Matrix had a liking for anagrammatic aliases. The hunch paid off. The letters spelled JOHN WORRELL KEELY, a Philadelphia carpenter whose late-nineteenth-century "Keely motor" was the most successful *perpetuum mobile* swindle in American history.

I stuffed some clothes into two cloth traveling bags and telephoned for a plane reservation. Shortly after four the next afternoon I stepped from the hot, humid air of Houston's Main Street into the cool lobby of the Shamrock. Frank Lloyd Wright is said to have remarked when he first entered this lobby: "Now I know what the inside of a jukebox looks like." For me, an ardent Oz buff, it was like walking into the palace of the Emerald City: thick green carpeting, garish green colors everywhere, including the kelly-green uniforms and the pillbox hats of the busy bellhops.

It was Dr. Matrix and Iva all right, although both pretended not to know me. "Yes, indeed," Dr. Matrix muttered through a bushy Karl Marx beard, his emerald eyes glittering with mild surprise. "I see your column now and then. Do you think *Scientific American* would be interested in an article on dynamaforce?" He ignored a guffaw from a stout man standing behind me who was wearing cowboy boots and a Stetson hat, long hair, sideburns, and a large steer's-head belt buckle with bright ruby eyes. I learned later that he was a prominent physicist from the Houston Space Center.

"Show Mr. Gardner our model, J.J.," said Dr. Matrix.

Iva led me by the hand into a small alcove behind the table where she and her father had been sitting. "Please don't give us away," she whispered. "And will you join us for dinner at seven?

The working model, submerged in a square, vertical tank of water about two feet on each side and ten feet high, was turning rapidly. Attached to a flexible and hollow belt of plastic were ten airtight cylindrical glass chambers, each holding a heavy rubber ball that was free to slide with the force of gravity (see Figure 20). The volume of air inside the belt and chambers as a whole obviously remained constant. Because the balls on the right compressed the air, sending it into the chambers on the left where falling balls created suction, it was clear that the chambers on the left displaced more liquid than those on the right. The result was a strong buoyant force on the left that caused the belt and the two wheels to spin perpetually clockwise.

"Isn't it sexy?" said Iva, grinning like the Cheshire cat. "All those balls moving up and down with the phalli." She told me that Mrs. Bloomfield Moore, widow of a Texas oilman, had already invested half a million dollars in the project. Dr. Matrix soon hoped to get the same amount from Miss Ima Hogg, a Houston socialite and daughter of a cattle and oil millionaire who was once governor of Texas.*

We dined in the Shamrock's Pavilion restaurant, which overlooks the hotel's enormous swimming pool. Iva spoke little while Dr. Matrix and I conversed.

"I like your touch," I said, "of putting UP and DN on the piston chambers. It reminds me of a gravity wheel C. L. Stong once pictured in his 'Amateur Scientist' department,

---

* It has often been asserted (by non-Texans) that Ima Hogg had a sister named Ura, but this is not true. James Stephen Hogg had no intention of being funny when he named his only daughter Ima. He was just proud of the fact that she was a Hogg. Miss Ima, as she was called in Texas, was greatly admired as a philanthropist and patron of the arts. She died in London in 1975 at the age of ninety-three.

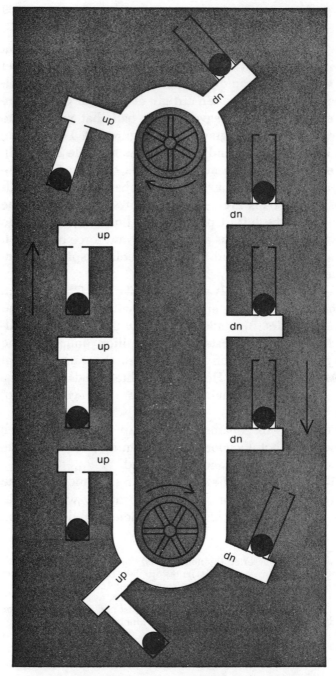

Figure 20. Dr. Matrix's dynamaforce generator

in *Scientific American*. The wheel had 9's on the down side. They outweighed the 6's they became on the up side."

"9 has remarkable inversion properties," said Dr. Matrix, nodding. "Mel Stover, a Winnipeg numerologist, has pointed out that the Roman numerals for 9 change to 11 when turned over, but 9 in the binary system—1001—doesn't change at all. It also stays the same if you write it like this."

Dr. Matrix removed a large gold pen from his striped seersucker jacket. With its felt tip he wrote on my note pad:

$$\mathcal{nine}$$

"Marvelous!" I exclaimed after turning the pad around.

"9 and 6, of course, are the digits that made 1961 the year that is the same upside down. You might try this on your readers. It was sent to me by Arthur Hall of Pinner in England. What invertible past year has the largest time lapse between itself and its inverse?"

"Let's see," I mumbled. "The difference between 1968 and 8961 is . . . hm . . . 6,993."

"You can do much better. The answer, by the way, is unique and naturally it turns out to be an important date in world history." (See Answers, Fifteen, I.) He paused while the waiter served us green tea. Three young men at a nearby table seemed to be studying the geometric pattern on Iva's hot pants.

"I was not surprised," Dr. Matrix went on, "to read in one of your columns about Arthur C. Clarke's denial that he intended HAL to stand for IBM." (HAL was the name of the spaceship computer in the film *2001: A Space Odys-*

sey. If each letter of HAL is moved forward one letter in the alphabet, it becomes IBM.)*

"Unintended word shifts are so commonplace," Dr. Matrix continued, "that they provide the strongest evidence we have for the soundness of Jung's concept of synchronicity. Consider the word OZ. L. Frank Baum, who started the Oz series of books, came from upper New York state. Shift each letter of NY forward once in the alphabet and you get OZ."

"By Glinda, so you do!"

"There's more," said Dr. Matrix. "Shift a second time, and OZ becomes PA, the abbreviation of Pennsylvania. Pennsylvania was the home state of Ruth Plumly Thompson, the young lady who continued the Oz series after Baum's death."

"I can't believe it," I said as I made my notes. "Speaking of *The Wizard of Oz,* does the word WIZARD have any numerological significance?"

"Everything does," sighed Dr. Matrix, while Iva's black eyes smiled at me over her teacup. "The word has a fantastic symmetry."

Dr. Matrix wrote down the alphabet and circled the first letter at each end, then the fourth letter from each end and then the ninth letter from each end (see Figure 21). They were the letters of WIZARD.

"Need I call your attention," he added, "to the fact that 1, 4, and 9 are the first three square numbers?"

"I think, my dear papa," Iva said, "that Gardner would like another problem his readers could work on."

---

* The earliest reference I could find on this astonishing word shift is in the periodical *IBM in Britain,* no. 47 (August 1968, p. 3), where its discovery is credited to John Roycroft, of London. In spite of the fact that the IBM logo is clearly visible on display terminals in *2001,* Clarke assures me that the word shift is entirely coincidental, and came as a great surprise to him.

Figure 21.   The alphabetical symmetry of WIZARD

Dr. Matrix tugged on his beard. "Have you noticed the high probability that a randomly selected pair of English words will have at least one letter in common? RED, ORANGE, YELLOW, GREEN, BLUE, PURPLE, and WHITE all share the letter *e*, but BLACK, the absence of color, does not. Peter Wexler of the University of Essex discovered that the English name for every integer, starting with 0, shares a letter with the next largest integer. 0 and 1 share *e*, 1 and 2 share *o*, 2 and 3 share *t*, and so on to infinity. There's a line in Longfellow's *Evangeline*, 'Warm by the forge within they watched the laboring bellows.' It contains all the letters in HENRY WADSWORTH LONGFELLOW."

"And the problem?" Iva asked impatiently.

It proved to be a large class of problems. The idea, Dr. Matrix explained, is to look for rare and special examples of what Los Angeles numerologist David L. Silverman calls a "heteroliteral" pair of names. These are names with *no* letter in common. Silverman found, for example, that only one of the fifty states is heteroliteral with respect to its capital. Readers might enjoy searching for it. (See Answers, Fifteen, II.) While doing so they may also discover which letter—there is only one—is not in the name of any state. (See Answers, Fifteen, III.)

What is so rare, Silverman once asked Dr. Matrix when the two met at a numerological convention in San Francisco, as a Friday in June? Silverman meant: What is the only other heteroliteral day-month pair? (See Answers, Fifteen, IV.)

"But back to numbers," said Dr. Matrix. "Perhaps your readers can determine the only letter that is not in the name of any number from 0 through 99, yet is in the name of every number greater than 99 and less than 1 million." (See Answers, Fifteen, V.)

Iva consented to an evening prowl through the muggy streets of downtown Houston. I thanked Dr. Matrix for having grabbed the tab, and while we pumped hands I said: "It's only fair to warn you that I'm wrestling with my conscience. Your dynamaforce—let's face it—is a colossal dynamafarce."

"Nevertheless," he replied gravely, "like Galileo's earth, the thing does move."

"I'm sure the local police would like to know who you really are."

"You have a point there," said Iva, pointing to my head.

"I couldn't possibly fail," said Dr. Matrix, "to disagree with you and Iva less."

It took me several blocks to figure out what that meant. I spent the night battling my superego. Fortunately no decision had to be made. Next morning, while I was having breakfast with Soph, Hooker and J.J. were seen entering their black Cadillac and speeding off without paying their hotel bill. The dynamaforce model was found to have a tiny battery-driven motor cleverly concealed inside its base.

Early that afternoon I carried my luggage through a jam of people, detectives, reporters, and photographers and stepped out of the Shamrock's air-conditioned lobby into a

sunbaked inferno. Heat waves shimmered over the spotless concrete. The traffic on Main Street was heavy, and
this brought to mind a letter from a reader, George E.
Mallinson, who pointed out that the initials of Charles A.
Reich, the author of *The Greening of America,* are
C.A.R.—that complicated nongreen mechanism whose
engine has become such a sooty albatross around the neck
of the world.

Low in the eastern sky, over the oil refineries and chem-
ical factories bordering the Houston Ship Channel, floated
a brownish haze, a baleful by-product of twentieth-century
seekers of black gold. It was a haze that would take more
than dynamaforce to dissipate.

16. Clairvoyance Test

An institute near Los Angeles (I must leave it nameless for the present) claimed in 1973 to be able to train anyone in the power of clairvoyance: the ability to perceive hidden or distant objects by extrasensory perception. Students were given tests before and after an intensive six-week training course—for which they paid a fee of $500. Their scores on the final test were invariably high.

A young lady who was a file clerk at the institute wrote me that the school's owner and founder looked suspiciously like Dr. Irving Joshua Matrix. And his chief assistant, she added, closely resembled Matrix's attractive Eurasian daughter, Iva. My informant managed to photocopy the clairvoyance test that students are given at the end of their training. The test is given below and the "targets" for each of the twenty-six problems are given upside down at the end of this chapter. Readers are urged not to look at the targets until they have completed the test.

Before starting the test you must have on hand the following material: a pencil and several sheets of paper, an unused matchbook, a deck of playing cards, a pair of dice, a ruler, scissors, a box of raisins, an eight-cent stamp, a penny, a nickel, a dime, a quarter, and a King James Bible. Answer each question as quickly as you can, without thinking too long or hard about what the target might be. Write your answers on a sheet of paper, and when you have finished the test, compare your answers with the given targets to see how many hits you have made. A score of more than fifteen hits indicates, according to the literature of the California school, an extremely high degree of clairvoyant ability.

1. Draw a circle around any one of the sixteen numbers in Figure 22. Cross out all cells in the same row and column. Draw a circle around one of the remaining nine numbers. Cross out the cells in its row and column. Circle one of the four remaining numbers. Cross out its row and column. Circle the single number that is left. Add the four circled numbers and write down the sum.

2. From an unused matchbook, which should have twenty matches, tear out any number of matches up to nine and discard them. Count the matches that remain.

1	2	3	4
5	6	7	8
9	10	11	12
13	14	15	16

Figure 22. A number test

Add the two digits of this number, then tear that many matches from the book and put them aside. Tear out two more matches. Write down the number of matches still in the matchbook.

3. From a deck of fifty-two playing cards remove the red queens, the black aces, the red fours, the six of clubs, and the jack of diamonds. Shuffle the deck, hold it face down and take the cards in pairs from the top. If the first pair contains a red and a black card, turn it face down and discard it. If both cards are red, put them face up on the table to start a pile of red pairs. If both cards are black, put them face up at another spot on the table to form a pile of black pairs. Continue through the deck in this fashion, discarding all red-black pairs and building up the piles of red-red and black-black pairs. When you finish, count the cards in each pile. Subtract the smaller number from the larger and write down the difference.

4. Draw a simple geometric figure. Inscribe within it another and different simple geometric figure.

5. Write the name of a wild beast.

6. Place two dice, A and B, on the table, any side up. Add the number on the top of A to the number on the bottom of B, then find the chapter of Genesis (in a King James Bible) that corresponds to the sum. Locate the verse indicated by the sum of the top of B and the bottom of A. Write the first word of that verse.

7. Think of any number, k, between 10 and 50. Place your finger on the bottom ESP symbol in Figure 23. Say "One." Tap the next symbol above it, saying "Two," and continue upward, counting aloud with each tap. When you come to the star, turn right and proceed counterclockwise around the circle, tapping and counting until you reach k. This may take you more than once around the circle. If it does, ignore the tail portion of the illustration.

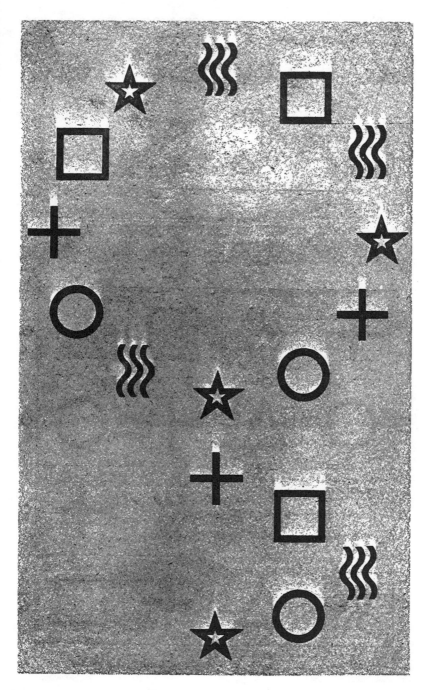

Figure 23. The *Q* test

After you tap the symbol on the count of k, stop and reverse direction, then tap and count from 1 to k as before, but this time go around the circle clockwise. The symbol you tapped for k should get the count of 1. (Do not make the mistake of starting the count on the symbol next to it.) Ignore the tail portion. Stop when you reach k again and write down the symbol just tapped.

8. Write a two-digit number between 10 and 50 that meets the following provisos: both digits must be odd and the digits must not be alike. (For example, 11 is ruled out because the same digit is repeated.)

9. Take any twenty cards from the deck and hold them face down. Turn the top pair of cards face up, leaving them on top of the packet, and cut the packet at any spot. Again reverse the top two cards and cut. Continue turning pairs and cutting for as long as you like. In reversing top pairs you will, of course, sometimes turn reversed cards face down again, but it does not matter. The procedure is designed to randomize the number of reversed cards in the packet.

Deal the randomized cards in a row on the table, taking care not to reverse any cards as you do so. Now turn over all the cards in even positions (2, 4, 6, . . . , 20). Count the number of cards that are face up in the row and write that number.

10. Form three equal piles of raisins on the table. There must be at least four raisins in each pile, and the number in each pile must be the sàme. Call the piles A, B, and C..

Take two raisins from A and put them in B.

Take three raisins from C and put them in B.

Count the raisins in A, then take the same number from B and put them in either A or C.

Take one raisin from either A or C and put it in B.

Write the number of raisins in B.

11. Think of any letter of the alphabet. Inspect the five columns in Figure 24 and circle that letter wherever it appears. Jot down the letter that is at the top of each column containing the letter you thought of. Convert these letters to numbers by using the cipher $A = 1$, $B = 2$, $C = 3$, and so on. Add the numbers. Using the same cipher key, convert the sum back to a letter. Write that letter.

12. Place one die on the left side of the table, the other on the right, each with any side up. Now obtain four products as follows: multiply the top two numbers, multiply the bottom two numbers, multiply the top of the left die by the bottom of the right, and multiply the top of the right by the bottom of the left. Add the four products and write down the sum.

A	B	D	H	P
C	C	E	I	Q
E	F	F	J	R
G	G	G	K	S
I	J	L	L	T
K	K	M	M	U
M	N	N	N	V
O	O	O	O	W
Q	R	T	X	X
S	S	U	Y	Y
U	V	V	Z	Z
W	W	W		
Y	Z			

Figure 24.   A letter test

13. Take a square sheet of paper, about eight inches on a side, and fold it in half four times so that the creases will mark a four-by-four matrix of small squares. Fold each crease both forward and back so that the paper will fold easily either way along every crease. Number the cells from 1 to 16 as shown in Figure 22.

Fold the square into a one-by-one packet, making each fold in any manner. Indeed, your folding may be as tricky as you please, including the tucking of folds between folds if you like. With a pair of scissors, trim away the four edges of the final packet so that it will contain sixteen separate squares. Spread the squares on a table. Some will be number side up, others number side down. Add all the face-up numbers and write the sum.

14. Write any number, provided only that its digits are not all alike. Form a new number by rearranging those digits any way you wish. Subtract the smaller number from the larger. Add the digits of the answer. If that gives a number of more than one digit, add the digits again, and continue in this way until a single digit remains. Increase that digit by 4 and write down the sum.

15. A pentagram, the mysterious occult symbol of medieval witchcraft and the ancient Pythagoreans, is shown in Figure 25. With a pencil put a dot anywhere you like inside the pentagonal border or on the border. Draw perpendiculars from the dot to each of the pentagon's five sides, extending the sides with a ruler if necessary. The perpendiculars are easily drawn by using the corner of a sheet of paper to provide a right angle. Add the lengths of the five perpendicular line segments by marking them along the edge of a sheet of paper. Measure the total length carefully with the ruler and write down its length to the nearest half inch.

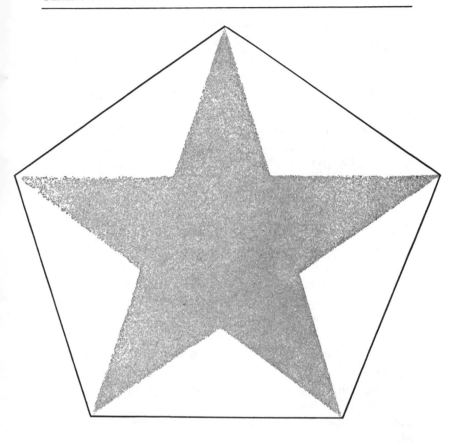

Figure 25.   The pentagram test

16. Write the name of a city that is the capital of a large foreign country.

17. Place in a row (in order from left to right) a penny, a dime, a nickel, an eight-cent stamp, and a quarter. Put a matchbook on any of the five objects. A move consists in transferring the matchbook to an adjacent object, either

left or right. Of course if the matchbook is at either end of the row, the next move will be limited to one direction. Move the matchbook randomly left and right as many times as indicated by the value in cents of the coin (or stamp) on which you first placed it. When you finish, if the matchbook is not on the penny, remove the penny from the row. Again move the matchbook as many times as indicated by the value of the object it is on. If the matchbook is not on the quarter when you finish, take away the quarter. Move the matchbook once. Write down the value of the coin (or stamp) on which it now rests.

18. A row of five face-up playing cards is formed on the table. From left to right the cards must be the nine of diamonds, the four of hearts, the queen of hearts, the ace of diamonds, and the seven of clubs. As you can see, there is one picture card, one ace, and one black card. Look the five cards over carefully, focus your attention on one of them, and then write down its name.

19. Think of a number from 1 through 16. Find that number on the border of Figure 26 and turn the page so that the number is right side up above the matrix. With the page still turned count the cells of the matrix beginning with the cell in the upper left corner until you reach your chosen number. Write down the ESP symbol inside that cell.

20. Write the name of a flower.

21. Shuffle a deck of cards. Assign to the face cards any value you wish from 1 through 10. (For example, you may decide to give each face card a value of 3.) Start dealing the cards face up from the top of the deck to form a pile on the table. Say "Ten" when you deal the first card, then continue with "Nine, eight, seven . . . ," counting backward as each card is dealt. If you put down a card that happens to coincide in value with the number named, stop

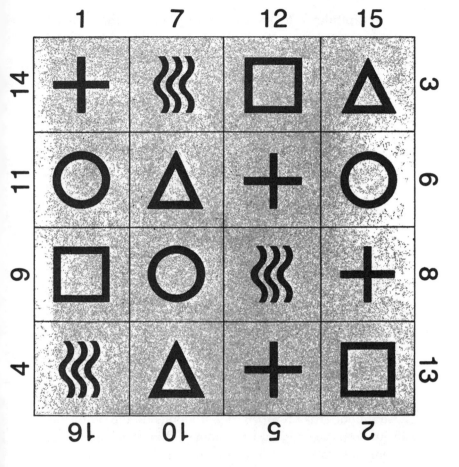

Figure 26. The rotating-matrix test

dealing on that pile and start another face-up pile next to it. If you fail to hit a coincidence by the time you deal and say "One," cover the last card dealt with a face-down card from the cards in your hand and begin another pile. Form four piles in this manner.

To recapitulate: As each face-up pile is being formed, count backward from 10 until you hit or until you count 1 without having made a hit. Each "failure" pile is covered with a face-down card. While you deal, be sure to keep in mind the value you assigned to the face cards. In our example it was 3; consequently, if you call "Three" when you deal a face card, it is counted as a hit and you start another pile.

After four piles have been formed add the values of your hits—that is, the face-up cards on the tops of piles. A face card has, of course, whatever value you assigned to it. Call the sum of your hits X. Discard X cards from those that have not been dealt. Count the cards left and write down the number.

22. Write a two-digit number between 50 and 100. Both digits must be even and not alike.

23. Roll a die on the table. Think of a number from 1 through 6 and put a second die on top of the first one so that the number you selected is on top of the stack. To your thought-of number add the sum of the two touching faces of the dice. Think of another number from 1 through 6 and add it to the previous total.

Remove the top die and turn it so that your second chosen number is on top. Place it alongside the other die. Lift up both dice and add the sum of their bottom faces to the previous total. Add 3 to the last result and write down the final sum.

24. Write the name of a color.

25. Put ten cards on the table. Turn them so that five are face up and five face down. Shuffle them around the table-top, mixing them thoroughly, then separate them into two sets, A and B. Reverse all the cards in set B by turning it upside down. Count the face-up cards in each set and write down the difference between the two numbers.

26. Select any digit from 1 through 5. Call it k. Look at the kth chapter of Revelation, in a King James Bible, and count to its kth word. Write the word.

(For sources of the twenty-six items see Answers, Sixteen.)

TARGETS FOR THE CLAIRVOYANCE TEST

1. 34.
2. 7.
3. 2.
4. Triangle and circle.
5. Lion.
6. And.
7. Star.
8. 37.
9. 10.
10. 8.
11. It is the letter you thought of.
12. 49.
13. 68.
14. 13.
15. 9¾ inches.
16. Paris.
17. 5 cents.
18. 4 of hearts.
19. Cross.
20. Rose.
21. 8.
22. 68.
23. 24.
24. Blue.
25. 0.
26. The.

17. Pyramid Lake

> Does the Great Pyramid of Cheops
> enshrine a lost science? Was this last
> remaining of the Seven Wonders of the
> World . . . designed by mysterious ar-
> chitects who had a deeper knowledge
> of the secrets of this universe than
> those who followed them?
>
> —Peter Tompkins,
> *Secrets of the Great Pyramid*

I was thumbing through one of those sleazy newsstand magazines devoted to the occult when a full-page adver- tisement caught my eye. It was a photograph of a six-foot- high transparent plastic model of the Great Pyramid of Cheops. Seated inside, clad only in sandals, was a beauti- ful dark-haired girl with oriental eyes. She looked exactly like Iva, the half-Japanese daughter of my old friend the renowned numerologist Dr. Irving Joshua Matrix.

I had seen advertisements before for small models of Cheops's pyramid (in the Edmund Scientific Company cat- alog, for instance) but not for models large enough to sit in. Each edge of the structure in the advertisement was mysteriously labeled with a different number from 1 through 10. There was no explanation of the numbering, nor could I find a price for the pyramid. However, $5 would buy an exact scale model six inches high. A booklet came with it, telling how the pyramid's "psi-org energy" would keep razor blades sharp, preserve rosebuds, and restore old typewriter ribbons. I would also be informed, said the advertisement, how to obtain the larger model, which was unconditionally guaranteed to cure my bodily ills, raise my intelligence, strengthen my psi powers, and build up my sexual potential. The address was Pyramid Power Laboratories, Post Office Box 123, Pyramid, Nevada.

Is there such a town as Pyramid, Nevada? I checked my atlas. Indeed there is. It is on the west shore of Pyramid Lake, about 35 miles north of Reno. I had no trouble obtaining the laboratory's telephone number from the Reno operator. A few minutes later I was talking with Iva herself.

"Come out and see us," she said. "Do you like to fish?"

I told her I did.

"Then bring a rod and reel. If the trout are under nineteen inches, you have to throw them back in the lake. The weather's marvelous here in May. Hot days, cool nights. Long time no see."

Before giving an account of my remarkable visit to Pyramid Lake, let me say something about "pyramid power." According to *Time* (Oct. 8, 1973), it all began seventy years ago when a French occultist, impressed by the excellent condition of mummies in the Great Pyramid, asked himself: Is it possible that the pyramid's shape does something peculiar to space and time? He put a dead cat inside a scale model of the pyramid. The body quickly dehydrated and mummified. Fifty years later Karel Drbal, a radio engineer in Prague, discovered that a razor blade, kept inside a six-inch-high model of the pyramid, never gets dull. More than that, a dull blade left inside for several weeks becomes sharp again! Drbal patented his Pyramid Razor-Blade Sharpener in 1959 (Czech patent no. 91304) and made a tidy fortune selling little cardboard and

Styrofoam models in Czechoslovakia. When Sheila Os-
trander and Lynn Schroeder reported this in their 1970
best seller *Psychic Discoveries Behind the Iron Curtain*, it
kicked off what *Time* called a "minicraze" in the U.S. and
Canada.

Max Toth's Toth Pyramid Company of Bellerose, New
York, sells a colored cardboard razor-blade sharpener.
In Glendale, California, G. Patrick (*G.P.* for Great Pyramid?) Flanagan sells a Cheops Pyramid Tent made of
vinyl. You sit inside it to improve your transcendental
meditation. Gloria Swanson, reported *Time*, sleeps with a
pyramid under her bed because it makes "every cell in my
body tingle." James Coburn likes to meditate in his pyramid tent.

Eric McLuhan, eldest son of Marshall McLuhan, has
been doing reasearch on pyramid power. A cover story
about him in the February 1973 issue of *enRoute* (a magazine published for Air Canada passengers) is based on an
interview with Eric when he was teaching "creative elec-
tronics" at Fanshawe College in London, Ontario. Eric
tells of putting one piece of hamburger in the center of his
Plexiglas pyramid and another on the floor of the pyramid.
Three months later the meat in the center was still fresh.
The other was unfit to eat. Eric believes the pyramid's
shape alters gravitational and magnetic fields inside and
above it. The dull edge of a razor blade is not improved,
he says, unless its edge is kept aligned on a north-south
magnetic axis. Blue steel blades work better than stainless
steel. John Rode, who runs an occult bookstore in Toronto,
experimented with larger pyramids. He has dehydrated
hundreds of eggs and ten pounds of porterhouse steak. "It
takes," Rode declares, "about twenty-three days to halt the
decaying process in meat. . . . We've fried the steaks and
eaten them. They're delicious."

Al G. Manning, head of the ESP Laboratory in Los

Angeles, has an article on pyramid power in the October 1973 issue of *Occult*. He recommends writing on a piece of paper a statement of something you intensely desire to happen. "Feed it with lots of love, and place it tenderly in the properly north-south oriented pyramid. The paper is then left in the pyramid for periods ranging from three to nine days with chanting and feeding of the thoughtform through the north side of the pyramid on at least a daily basis." Manning reports spectacular results.

The 1974 issue of *Other Dimensions,* an annual review of the paranormal, has an informative article on "The Lady Who Lived in a Pyramid." It seems that when Tenny Hale, an Oregon sensitive, sat inside her large wood model of the Great Pyramid, she found that her psychic powers were so enhanced she emerged and typed one hundred different prophecies. She also reports that when she waters plants with water stored under her pyramid, the plants grow four times as fast as plants watered with ordinary water.

James Mullin, research director of the Southern Parapsychology Foundation in San Diego, is quoted as warning people not to spend too much time in a pyramid. Pyramid energy kills bacteria that cause decay. Since bacteria can be good as well as harmful, says Mullin, prolonged exposure to pyramid power may be a health hazard.

Readers who wish to conduct their own experiments will find it is easy to make a cardboard model of a pyramid with a square base. Simply cut four cardboard triangles and tape their sides together. Each triangle should have its altitude in golden ratio to half its base. Herodotus was the first to suggest that the area of each face of the Great Pyramid is equal to the square of the pyramid's height. Let x be the "apothem" (altitude of a face) and the base equal to 2. The area of each face is x. The apothem is the hypotenuse of a right triangle whose legs are 1 (halfway across the pyramid base) and the height of the pyramid. Applying the Pythagorean theorem, we find that the pyramid's height is $\sqrt{x^2-1}$. If the area of the face equals the square of the height, we have the equality $x^2-x-1=0$. This gives x a value of $\frac{1}{2}(1+\sqrt{5})$, which is phi, the golden ratio, or 1.6180339887. . . . In other words, if the pyramid's base is 2, its apothem is phi and its height is 1.2720196495 . . . , the square root of phi. There is a surprising bonus. Divide 4 (twice the base) by the square root of phi (height) and you get 3.1446 . . . , a close approximation of pi (see Figure 27).

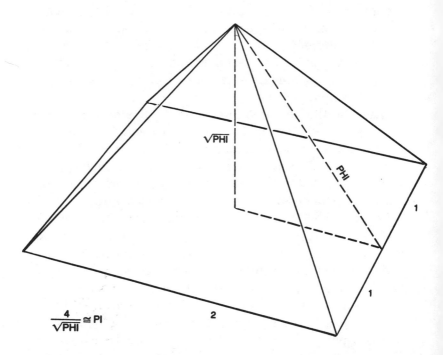

Figure 27.   The perfect phi pyramid, after Herodotus

The difference in slope between a perfect pi pyramid and a perfect phi one is close to one minute of arc: the slope is about 51 degrees 50 minutes for phi and about 51 degrees 51 minutes for pi. The variation in slope is too minute to be detectable on a small model. The Great Pyramid itself is now so irregular that all one can say for certain is that its slope is close to 52 degrees. No one knows if the Egyptians intended to embody in the pyramid's shape pi or phi, or both or neither.

But back to my trip west. I arrived in Reno by plane on the evening of Tuesday, May 28—a perfect day and a perfect number. I spent the night in a Reno hotel. Early next morning I rented a car, drove east to Sparks, then north on Route 33 to Sutcliffe, on the western rim of Pyramid Lake. I had been there once before. My father, a geologist, had taken me to Pyramid Lake when I was a boy. I remember him telling me how the lake was the shrinking remnant of a vast prehistoric body of water called Lake Lahontan that had covered most of the northwestern part of Nevada during the Pleistocene ice ages.* He explained how the lake was fed by waters from the Sierra Nevada, carried mainly by the Truckee River, which flows through Reno. He took me by rowboat to some of the cone-shaped islands that give the lake its name, particularly to Pyramid Rock on the lake's eastern bank. I own my father's copy of the journal kept by John Charles Frémont, the famous explorer and

---

* The slow shrinking of Pyramid Lake has been greatly accelerated in recent years by diversion of its water for nearby irrigation projects, and by urban development. The lake has become so alkaline that the Lahontan trout, once a basic food of the Indians, have almost vanished from the water. As one writer has put it, the greening of the desert is spelling the death of the lake.

The native Indians are also in a lamentable state of decline. Poverty and unemployment are high. It was, of course, the willingness of the Indians to work for low wages that prompted Dr. Matrix to locate his factory in the area.

Figure 28.   Pyramid Rock in Pyramid Lake, Nevada

politician (he was once governor of California) who discovered the lake in 1844. My father had marked the passage
in which Frémont describes the 300-foot-high calcareous
formation called Pyramid Rock: "We encamped along the
shore, opposite a very remarkable rock . . . which from
the point we viewed it, presented a pretty exact outline of
the Great Pyramid of Cheops." (See Figure 28.)

I stopped at Sutcliffe to buy a fishing license. Pyramid
Lake is entirely within the huge Pyramid Lake Indian

Reservation owned by the Paiute Indians. One is not allowed to fish without a permit obtained from tribal officers. After getting my permit I drove north on the deserted dusty road that winds along the lake's barren western shore.

The day was warm and cloudless. Through the right window of my car I could see, beyond the sagebrush, the deep Prussian blue of the lake. Jagged spires and pinnacles along the opposite shore were casting purple shadows over the water, and above the turrets the Lake Range mountains undulated in soft shades of green and pink. Just before entering Pyramid I turned off on a side road, as Iva had instructed. The road led into a beautiful hidden canyon, and soon ended at a massive factory building of steel and concrete. It had been built in the shape of the Great Pyramid of Cheops. Large blood-red numerals gleamed on the structure's edges.

A pudgy Paiute opened the front door for me. He grinned broadly, revealing a mouth that contained only a single front tooth. (His name was Ree, I later learned. Everyone called him One-Tooth Ree.) Down the hall and walking toward me came Iva and her father. Iva was wearing bright orange pants, a beaded Indian headband, and a charm bracelet of little silver pyramids that jangled pleasantly as she walked. We embraced. Dr. Matrix stood by, tall and bony, his canny green eyes glittering behind rimless pentagonal spectacles.

They took me on a quick tour of the factory. In one wing about twenty Indians were assembling the six-inch pyramid models. In another wing a smaller group of Indians were cutting and packaging the unassembled sides of the larger model. Iva excused herself, and I followed Dr. Matrix up a helical stairway to his office in the factory's apex.

Psi-org, Dr. Matrix explained, leaning back in his desk

chair and touching fingertips to fingertips, combines abbreviations for psychic and orgone energy. They are different names for the same force. The psi field, which produces the human aura and is reponsible for all psychic
powers, is none other than what Wilhelm Reich, Freud's
controversial Austrian disciple, called orgone energy.

"I remember Reich's orgone," I said. "It comes from
outer space. It makes the stars twinkle, the sky blue, and
Orson Bean happy." *

"Precisely," said Dr. Matrix. "The hundreds of pyramid
islands in Pyramid Lake trap the energy and this gives the
strong blue color to the water. Reich's great discovery, as
you know, was that orgone could be accumulated by building a box with wood on the outside and sheet iron on the
inside. The organic material lets orgone through; the metal
interior reflects. What I call the 'bluehouse effect' takes
over. Abnormally high concentrations of psi-org energy
build up inside the box. Reich's basic idea was sound, but
he had the shape of his box wrong. The Egyptians knew
all about psi-org energy. They used it, you know, to float
heavy stones across the desert when they built their pyramids. They were the first to discover that the shape of the
Great Pyramid concentrated psi-org. I was the first to discover that if this shape is combined with Reich's principle
of laminated substances, the bluehouse effect increases by
a factor of 777. I call it the pi-phi-psi pyramid."

"But your pyramids aren't laminated," I said. "They're
just single sheets of Plexiglas."

"Wrong," said Dr. Matrix. "Examine them more care-

---

* I had recently read Orson Bean's naive little book *Me and the Orgone*
(New York: Fawcett paperback, 1972). For a good recent summary of
Reich's orgone energy claims (as funny as any of Orson's comedy rou-
tines), see David Boadella's paperback *Wilhelm Reich* (New York:
Dell, 1975).

fully and you'll see that every side consists of two thin
layers of plastic. Each sheet is made with a different and
secret formula. The outer layer transmits psi-org, the inner
layer reflects it."

"Has this been verified?"

"To the hilt. We've done hundreds of carefully controlled tests under the supervision of Dr. Harald Puton, a
very competent Belgian physicist who was active in the
Scientology movement in Brussels a few years ago. He
found that every form of psi energy is increased by sitting
under a pi-phi-psi pyramid. One is more telepathic, more
clairvoyant, more precognitive. It is easier to initiate out-
of-body experiences. Uri Geller, the Israeli psychic, visited us a few weeks ago and found that when he was inside the pyramid, any metal object he touched melted instantly. Psychic healing is enormously accelerated. Last
week One-Tooth brought in his sister. She'd broken her
left leg. After one hour in a pyramid her leg was as good
as new."

Was Dr. Matrix smiling faintly? It is always hard to
know where his beliefs end and deception begins. The
body's aura, he continued, is more intense inside a pyramid. He took from his desk two Kirlian photographs of a
living butterfly. The one made outside the pyramid
showed only a faint white aura. The one made inside
showed a bright blue aura that extended several inches
beyond the butterfly's wings.

"Would you mind explaining the significance of these
numbers?" I asked, pointing to the scarlet numerals on a
pyramid paperweight.

Dr. Matrix did not mind. Each of the pyramid's eight
edges, he said, bears a different number in the set of
integers from 1 through 10. At each vertex (the four base
corners and the apex) the sum of all edges meeting at that

vertex is $4^2 = 16$. I picked up the paperweight and turned it in my hands. What he said was true. At each base corner the numbers on the three meeting edges added up to 16. The four edges meeting at the top also added up to 16.

The labeling, Dr. Matrix went on, makes it a magic pi-phi-psi pyramid. The magic increases the pyramid's power. He called my attention to the fact that p, the initial letter of PYRAMID, is the sixteenth letter of the alphabet. Between the letters pi in PYRAMID is the name of the mother of Jesus spelled backward.

"What about the d?"

"It's the alphabet's fourth letter. It symbolizes the structure's four sides and the square root of 16, the magic constant. You might ask your readers to see if they can number a pyramid's edges properly to make it magic. There's only one way to do it. Of course we don't count rotations and reflections as being different."

"Excellent," I said, sketching the pyramid on my note pad and numbering the edges in case I forgot the solution. (See Answers, Seventeen.) "Anything else you can tell me? Anything numerologically interesting about Nixon and Watergate?"

"*Everything* is numerologically interesting," he said. "But Nixon's future is too painful to discuss. Incidentally, there's a town named Nixon just south of the lake. He could do worse than retire there. A few hours a day sitting on the summit of Pyramid Rock would do the president a world of good. It would certainly be less uncomfortable than his present seat on top of the country's pyramidal power structure."

"Any comments on Henry Kissinger? Someone told me recently that if that celebrated Polish opera star Wanda Waleska married Howard Hughes, divorced him and married Henry, she'd be Wanda Hughes Kissinger now."

"Your humor," said Dr. Matrix without smiling, "underwhelms me. You're familiar, of course, with the last verse of chapter 13 in the last book of the Bible?"

I nodded. "Let him that hath understanding count the number of the beast: for it is the number of a man; and his number *is* Six hundred threescore *and* six."

"Well, I don't want to suggest that Kissinger is the Beast, but here's a little number curiosity your fans may find amusing. Let *a* equal 6, *b* equal 12, *c* equal 18, and so on, taking the multiples of 6 in sequence. Then add the values of KISSINGER. The total is 666." *

"Marvelous," I chuckled, scribbling furiously. "Yesterday while I was on the plane it occurred to me that NEVADA, which has six letters, was admitted as the 36th state. And 36 is the square of 6."

"And dice," said Dr. Matrix, "have six sides. 6 is the highest number on a die and 36 is the highest number on a roulette wheel—two reasons why the 36th state contains the nation's three largest gambling resorts. Six is the sum of 1, 2, 3, and 36 is the sum of the cubes of 1, 2, 3. Of course 6 is also the product of 1, 2, 3. Did you know, by the way, that 1, 2, 3 is a unique set of distinct positive integers, relatively prime to each other, such that the sum of any two is divisible by the third?" †

"No, I didn't," I said as I wrote this down.

---

* First noted by Wayne L. Johnson, of Albany, Oregon, and published in his letter to the *News-Register*, McMinnville, Oregon (March 2, 1974). Robert A. De Forest of Sheridan, Oregon, called Dr. Matrix's attention to this letter.

† For a quick proof of this, together with a proof that 2, 3, 5 is a unique set of distinct positive integers such that the product of any two leaves a remainder of 1 when divided by the third, see "Two Unique Sets of Numbers," an answer by E. P. Starke to problem E1234 (note the problem's number!), *American Mathematical Monthly* 64 (1957):275.

"It is also noteworthy," Dr. Matrix continued, "that 666 is the sum of all the numbers on a roulette wheel. It is the thirty-sixth triangular number, and there are just *six* triangular numbers made up from a single digit."

"Are you including the single-digit triangles, 1, 3, and 6?"

"Naturally."

"And the other three?"

"55, 66, and 666." *

"I once suggested somewhere," I said, "that if the devil plays rotation pool, he plays it on a huge table with 666 balls in triangular formation before the break."

"Quite likely. 6 is the devil's number, all right. It was the sixth hour of the sixth day of creation when he tempted Eve. But it was the Egyptian sensitives who made the fullest use of the square of the sum of 1, 2, 3."

One-Tooth stuck his head in the doorway. "Did you call?"

Dr. Matrix shook his head and waved Ree away. "As I was saying, Osiris had 36 forms. The human body was divided by the Egyptians into 36 parts, each subject to a different demon, each controlled by one of the 36 parts of the Egyptian zodiac. And I could bore you for hours with a discussion of Nabokov's use of 36 in *The Real Life of Sebastian Knight*. Sebastian's house number is 36. His hospital room number is 36. He dies in 1936 at the age of 36,

---

* E. B. Escott, in 1905, showed that 1, 3, 6, 55, 66, and 666 are the only rep-digit triangles with fewer than thirty digits. (See L. E. Dickson, *History of the Theory of Numbers*, 2:33.) In 1972 David W. Ballew and Ronald C. Weger announced that their computer program proved that no other such numbers exist. (See *Notices of the American Mathematical Society* 19 (1972):A-511.) A number theoretic proof is given by Ballew and Weger in "Repdigit Triangular Numbers," *Journal of Recreational Mathematics*, 8 (1975–6):96-98.

and so on. But," and Dr. Matrix glanced at his digital wristwatch, "my time is limited."

"Is it necessary to be nude when you sit inside a pyramid?" I asked.

"No, but it helps. We have several opaque models on the beach for visitors too modest to use transparent ones. Last week Judy Clutch, a schoolteacher from Wadsworth, was inside one of them for only five minutes before she was so blue with psi-org that she leaped outside, ran all the way to Pyramid, and streaked down the main street until the sheriff caught her and took her to lunch."

I asked about the pyramid's power to prevent and even reverse the decay of organic matter. Dr. Matrix told me he had applied for numerous patents that exploited this aspect of psi-org: a pyramid refrigerator and freezer, a pyramid coffin (no embalming necessary), and a pyramid septic tank. He was even experimenting with a pyramid outhouse that kept one "regular" in addition to dehydrating and purifying the waste. Dr. Matrix called it the "pi-phi-psi-shi house."

I stayed five days in Reno. Mornings and afternoons I fished in Pyramid Lake for king-size trout and a tasty variety of lakesucker that the Indians call *cui-ui*. Iva joined me evenings for dinner at one of the resort lodges near Sutcliffe. On Saturday I rented a motorboat and we visited Anaho Island, a 250-acre bird sanctuary where we watched the white pelicans go through their spring mating ritual. Iva's large straw hat was shaped like the Great Pyramid. She said it made her think better and corrected her astigmatism.

Several days after I had returned to Manhattan I was distressed to see in the *New York Times* that Dr. Matrix was about to be indicted by the state of Nevada. It seems he had been selling pyramid franchises for several thousand dollars each to a large number of people in other states, who in turn had been trained to sell similar franchises to other distributors, and so on. It was a typical pyramid scheme of the sort that sooner or later is bound to collapse.

I tried to reach Iva, but the laboratory phone was no longer a working number. Two days later I read in the *Times* that when state troopers went to the factory to arrest Dr. Matrix, they found no one there except One-Tooth Ree. He had just finished burning all the factory records in a large outdoor bonfire. He handed the police a letter from Dr. Matrix. It explained that on June 6 at 6:00 A.M.—the sixth hour of the sixth day of the sixth month—Dr. Matrix and his daughter had entered one of their large pyramids and teleported themselves to a monastery in Tibet.

18. The King James Bible

When I turned on the mechanism that records incoming telephone calls (I had been out of town), I was surprised to hear Iva's familiar voice. Her message was cryptic: "Isaiah 50, verse 2, sentence 1." This was followed by a telephone number. I checked the reference in my King James Version and read: "Wherefore, when I came, *was there* no man? when I called, *was there* none to answer?"

I returned the call and listened incredulously while Iva told me about her father's latest exploit. It seems that the monastery in Tibet, where the two had been living for more than a year, was not (as I had supposed) Buddhist. It was run by an obscure American sect of Protestant Fundamentalists called the Church of the True Word. For six months Dr. Matrix had occupied himself by writing a thirteen-volume commentary on the King James Version.

This massive set of books, Iva informed me, had been privately printed in Switzerland, in English, and translations were now under way for French, German, and Russian editions. Her father was supervising the translations and handling the European sales of the English edition from his present residence in Paris. She was in New York for a conference with a major publisher, which had expressed interest in taking over the set's distribution in the U.S.

Dr. Matrix wanted me to have a complimentary copy of his great work, and of course I was almost as happy to accept it as I was to see Iva again. Unfortunately it is impossible to give an adequate review of this monumental commentary here. One would sooner review Mortimer J. Adler's new edition of the *Encyclopaedia Britannica*. Nevertheless, I shall do my best.

The set's title is *The King James Bible, with Commentary and Critical Notes by Irving Joshua Matrix, D.N.* The Apocrypha are not included. Each volume is eleven by eight inches, two and a half inches thick, and weighs four pounds. The paper is thin but of excellent quality. Throughout the work every cardinal number mentioned in the text proper is printed in green and every ordinal number is in blue. Passages referring in any way to mathematics appear in dark maroon ink. Passages containing material of combinatorial interest, with respect to letters or words, are printed in purple.

A quotation from Job 14:16, "Thou numberest my steps," appears on the title page. The ten volumes devoted to the Old Testament are preceded by the quotation "Teach *us* to number our days (Ps. 90:12). "The very hairs of your head are all numbered" (Matt. 10:30) introduces the three volumes on the New Testament.

Dr. Matrix's commentary has little in common with the kind of numerology practiced by the cabalists of old Judea (who assigned numbers to each of the 22 letters in the Hebrew alphabet) or by Leonard Bernstein (who adopted the same technique when he composed the score for Jerome Robbins's 1974 ballet *The Dybbuk*). Nor is Dr. Matrix concerned with the medieval Christian technique of number juggling based on the numerical values of Greek letters. Occasionally Dr. Matrix refers to this enormous literature of Hebrew and Christian number mysticism, but for the most part he is concerned solely with the King James text, exploring its combinatorial patterns from the standpoint of the modern kind of numerology that he himself pioneered.

Consider Dr. Matrix's observation, in his introduction, on the number of books in the Old Testament and the New Testament. OLD has three letters, TESTAMENT has nine. Put the two digits side by side and you have 39, the number of Old Testament books. Similarly, NEW has three letters. The product of 3 and 9 is 27, the number of New Testament books.

Dr. Matrix's long note on Genesis 4:14–15 will surprise many readers. This is the passage where Cain expresses his fear that everyone he meets will want to slay him. Since Cain and Abel were the first-born of Adam and Eve, the world's population must have consisted of no more than Cain and his parents and perhaps a few sisters. Who, then, would have tried to kill him?

Dr. Matrix finds the answer in St. Augustine's *The City of God,* book 15 (a work, by the way, that Augustine divided, says Dr. Matrix, into twenty-two parts to parallel the twenty-two letters of the Hebrew alphabet). We know from the chronology supplied in Genesis 4 and 5 that Abel was murdered about the year 129 after creation, and we know from 5:4 that Adam "begat sons and daughters." Is it not reasonable, asks Dr. Matrix, to assume that Adam and Eve had a child each year? By the year 129 there would have been 129 offspring. Assume that the sexes were equally divided, that there were no deaths before Abel was murdered and that brothers and sisters married and had one child per year from the age of, say, eighteen on. It is easy to calculate that at the time Cain killed his brother, Adam and Eve would have had more than 3,000 grandchildren and more than 90,000 great-grandchildren. Adding to this the great^2-, great^3- and great^4-grandchildren, Dr. Matrix estimates that the world's population at the time of the murder was about half a million. (See Figure 29.)

On the ages of the patriarchs, Dr. Matrix has many curious observations. Methuselah's age of 969 (Gen. 5:27) is, obviously, a palindrome. It can be expressed as the difference between the squares of two numbers, each less than 500, in just four ways. Turn 969 upside down and you get the palindrome 696, which also can be expressed as a difference between two squares, with roots less than 500, in just four ways. Moreover, 969 is the 17th tetrahedral number. With 969 cannonballs one could build a tetrahedral pyramid having 17 balls to the edge. Methuselah's father, Enoch, also had an unusual life span. He lived 365 years (Gen. 5:23), or one year for each day in a year before he was translated (Gen. 5:24; Heb. 11:5).

Dr. Matrix proves that Methuselah died in the very year of the Flood. (See Figure 30.) He was 187 when Lamech was born (Gen. 5:25), Lamech was 182 when Noah was born (5:28), and Noah was 600 when the Flood started (7:6). The sum of 187, 182, and 600 is 969. "Certain it is that Methuselah did not survive the flood, but died in the very year it occurred," wrote Augustine (*The City of God*, book 15, section 11).

Noah's father died five years before the Flood. Did Noah permit his grandfather, Methuselah, to perish in the Flood, as H. S. M. Coxeter suggests in a note on "An Ancient Tragedy," *Mathematical Gazette*, 55 (1971):312? Countering this view, Dr. Matrix cites the opinion of Rabbi Solomon Itzhaki, better known as Rashi, an acronym based on the Hebrew initials of his name. Rashi was a twelfth-century French rabbinic commentator on the Bible and Talmud. Methuselah, Rashi argued, died just before the Flood. God waited until the seven days of mourning

Figure 29. What was the population of the world when Cain
killed Abel? (Engraving by Doré)

Figure 30. Did Methuselah die in the Flood? (Engraving by Gustave Doré)

had ended ("For yet seven days, and I will cause it to rain
. . ." Gen. 7:4) before starting the Flood.

From the thousands of notes Dr. Matrix has on individ-
ual numbers I shall mention only a few that caught my attention. The "two hundred thousand thousand"
(200,000,000) horsemen in Revelation 9:16 is identified as
the largest integer explicitly mentioned in the Bible. The
first integer to be mentioned is also the smallest, the ordinal 1 ("first day") of Genesis 1:5. The smallest integer
not mentioned in the Bible, Dr. Matrix asserts in the same
note, is 43.

On the 153 fishes caught in the unbroken net (John
21:11) Dr. Matrix first repeats Augustine's analysis in *Trac-
tates on the Gospel of Saint John*. We take 10 (ten commandments) as a symbol of the old dispensation and 7
(seven gifts of the spirit) as a symbol of the new. They add
to 17, and the sum of the integers 1 through 17 is 153. Dr.
Matrix points out that 17 is the seventh (holiest) prime.
When 153 is written in binary (10011001), it is palindromic.* On the 276 shipwrecked souls (Acts 27:37) Dr. Matrix observes that 276 is the sum of the fifth powers of 1, 2,
and 3.

The number 490—the "seventy times seven" that Jesus
told Peter was the number of times one should forgive a
brother's sins (Matt. 18:22)—is, Dr. Matrix tells us, the
number of ways 19 can be partitioned into positive integers that add to 19. It appears in Daniel 9:24 as "seventy
weeks." If one exceeds 490 sins by committing a 491st (the
basis of Vilgot Sjoman's 1965 film titled *491*), the ordinal
number of that unforgivable sin is a prime. "Moreover,"
adds Dr. Matrix, "it is the sum of the squares of the primes

---

* For more on 153, see chapter 3.

3, 11, and 19, and is the difference between the squares of consecutive numbers 245 and 246."

I had not known until I looked over Dr. Matrix's commentary that 666, the number of the beast in Revelation 13:18, had appeared twice earlier in the Old Testament: in 1 Kings 10:14 as the number of gold talents given to Solomon in one year and in Ezra 2:13 as the number of children of Adonikam, a name meaning "lord of enemies." Nor had I realized that Revelation is the 66th book of the Bible and that 18, the verse that mentions 666, is the sum of the digits that compose 666. Since Dr. Matrix devotes more than fifty pages to this history of interpretations of 666, I shall make no attempt here to summarize his comments. One paragraph, however, struck me as being of unusual interest. Medieval numerologists made much of the number 1,480, obtained by summing the values assigned to the Greek letters of CHRISTOS. Dr. Matrix describes a curious way in which this number is linked to 666. Construct a square of side 1,480. Its diagonal (ignoring the fraction) is 2,093. Now construct a circle with a circumference of 2,093. The circle's diameter, again ignoring the fraction, is 666!

Another curious linkage, between 666 and pi, is provided by a splendid cryptarithm that Dr. Matrix credits to an expert composer of such puzzles, Alan Wayne of Holiday, Florida:

$$SIX + SIX + SIX = NINE + NINE$$

Each letter represents only one digit, and different letters represent different digits. The solution is unique. If the reader can find it, he will see two ways in which pi creeps into the pattern. (See Answers, Eighteen, I.)

The biblical approximation of pi is given in 1 Kings 7:23 and is repeated in 2 Chronicles 4:2. Both verses speak of a circular "molten sea" with a diameter of ten cubits and a circumference of thirty. One might suppose the Old Testament writers had no better estimate of pi than 3, but Dr. Matrix thinks otherwise. Consider the verse in which pi is first mentioned, 1 Kings 7:23. The initial 1, subtracted from the terminal 23, gives the ratio 7:22, and 22 divided by 7 is 3.14+, a fair approximation of pi. For a still better value, twice 7 is 14, half of 2 is 1, and twice 3 is 6. This gives 1416, the first four decimals of an excellent approximation of pi.

Dr. Matrix reminds his readers that on the basis of the third book of the Old Testament, 14th chapter, 16th verse, he had predicted in 1966 (see chapter 8) that the millionth digit of pi would prove to be 5. This was verified in 1974, when pi was computed in Paris to a million decimal digits. (The millionth *decimal* digit, excluding the initial 3, is 1.) The Bible's poor estimate of pi, says Dr. Matrix, may be due to a temporary destruction of the "tables" by Moses (Exod. 32:19).

The most mysterious number in modern physics, the fine-structure constant, is not overlooked by Dr. Matrix. If the elementary charge e is squared and is divided by the product of c (velocity of light) and h (the quantum constant), the result is 1/137, the reciprocal of the fine-structure constant. Dr. Matrix finds 137 first mentioned in the Bible as the age of Ishmael (Gen. 25:17), later as the age of Levi (Exod. 6:16) and the age of Amram (Exod. 6:20). Dr. Matrix apologizes for not giving his own theory of how 137 can be derived from these three names and how the names are connected with relativity theory and quantum mechanics. (He says he is writing a monograph on it.) He does, however, repeat the anecdote about how the physicist Wolfgang Pauli, after he had died, asked God why he had picked 137. God handed him some papers covered with formulas and said, "It's all here." Pauli studied the formulas carefully, frowned, looked up, and said, *Das ist falsch.*

There are three passages in Revelation (7:4, 14:1, and 14:3) that speak of 144,000 redeemed saints standing before God's throne, singing a new song, with God's name written on their foreheads. As Revelation 7 makes clear, they represent 12,000 saints from each of the 12 tribes of Israel (see also Matt. 19:28). Dr. Matrix devotes several pages to the history of how this has been interpreted by biblical commentators, from Origen to such contemporary Adventist sects as Jehovah's Witnesses and the Seventh-Day Adventists. The Witnesses believe an invisible resurrection began in 1918 and that exactly 144,000 saints eventually will go to heaven in contrast to millions who will never die on earth. Seventh-Day Adventists believe that 144,000 *living* saints will be translated to heaven on the day of the Second Coming.

Drawing on his extensive knowledge of early Adventist literature, Dr. Matrix quotes an amusing passage from an early prophetic vision of Mrs. Ellen Gould White, the remarkable woman who founded the Adventist movement. Mrs. White, in a trance, saw the 144,000 saints standing on a sea of glass in "a perfect square." She failed to realize, writes Dr. Matrix, that the square root of 144,000 is not 120 or 1,200 but the irrational number 379.4733+.

Years later, Dr. Matrix tells us, when Mrs. White wrote her most famous book, *The Great Controversy Between Christ and Satan,* she described the totality of the redeemed as standing "in a hollow square, with Jesus in the midst." In later editions this became "The glittering ranks are drawn up in the form of a hollow square about their King, whose form rises in majesty above saint and angel."

Perhaps, Dr. Matrix reasons, the subset of 144,000 translated saints also stand in a "hollow square." He cites an analysis made by Harold F. Williams, who teaches mathematics at the Adventist-affiliated Platte Valley Academy in Shelton, Nebraska. Assuming that the saints are in a perfect square array, with a central square hole, its sides parallel to the sides of the outer square, there are just thirty-six possible ways such a hollow square can be formed. Note that 36 is both a square and a factor of 144,000. Can the reader give the side of the smallest hollow square for 144,000 saints? (See Answers, Eighteen, II.)

Although Dr. Matrix makes no mention of Hal Lindsey's best-seller *There's a New World Coming* (New York: Bantam, 1973), I think it may be amusing to add that according to Lindsey's commentary on the Apocalypse the 144,000 saints will be converted Jewish evangelists who will preach the gospel during the world's seven years of great tribulation. Lindsey is now second only to Herbert Armstrong and Garner Ted Armstrong as the country's leading explicator of biblical prophecy. He currently teaches at a Fundamentalist training center in California called the J. C. Light and Power Company.

One of the most amazing of all the number patterns found in the Bible by Dr. Matrix has to do with the number of chapters in the four gospels. Taking them in sequence, Matthew, Mark, Luke, and John have respectively 28, 16, 24, and 21 chapters. Keeping this sequence, we form the complex fraction $(28/16)/(24/21)$. Now reverse the order of these four numbers to get $(21/24)/(16/28)$. Each common fraction has been changed to its reciprocal, and the numerator and denominator of the complex fraction have also switched places. In spite of this transformation, believe it or not, the value of the entire expression remains unchanged. It is 1.53125. The first three digits, notes Dr. Matrix, are the number of fishes in the unbroken net. The last three decimal digits are obtained from the first two (5, 3) by raising 5 to the power of 3.

To appreciate how remarkable it is that a complex fraction of four numbers would remain the same after this radical kind of transformation, the reader is asked to search for four other integers, all different, that will form a complex fraction having the same property. (See Answers, Eighteen, III.)

Revelation 7:9 speaks of the total number of the saved as "a great multitude, which no man could number." If no man can number them, reasons Dr. Matrix, then the saved must form an uncountable infinite set. Since the number of human beings that will have lived on the earth must necessarily be a finite number, we must conclude that an uncountable number of planets in the universe bear intelligent life. If the number of such planets were countable, the number of souls on all of them, at any given moment of cosmic history, would also be countable, thus contradicting Revelation.

Dr. Matrix finds a lively interest in mathematics throughout the Old Testament. Acknowledging his debt to G. J. S. Ross, a mathematician at the University of Cambridge, Dr. Matrix points out that "men began to multiply" (Gen. 6:1) as soon as they were created and continued to do so in New Testament times (2 Pet. 1:2; 2 Cor. 9:10). They performed division (Gen. 15:10; Num. 31:27), addition (2 Pet. 1:5), and subtraction (Gen. 18:28). They learned how to extract "the roots thereof" (Ezek. 17:9) and how to wrestle "against powers" (Eph. 6:12).

As for geometry, great rulers were brought down (Ps. 136:17), and from Syracuse they "fetched a compass" (Acts 28:13). Noah constructed an ark. The ancient Hebrews were familiar with "axes" (1 Sam. 13:21), and David refused to accept something until he had "proved *it*" (1 Sam. 17:39). Paul knew all about four-dimensional geometry (Eph. 3:18), and Joshua continued an ark along a Jordan path (Josh. 3).

Abstract algebra also was not unknown to the Patriarchs. Opening a matrix is mentioned several times (Exod. 13:12, 15; 34:19; Num. 3:12, 18:15). Ezekiel thought "rings" were "dreadful" (Ezek. 1:18). Jeremiah complained of "abominations" in "the fields" (Jer. 13:27). Peter was troubled by "four quaternions" (Acts 12:4), and Jesus had a low opinion of those who "seeketh after a sign" (Matt. 16:4).

Dr. Matrix credits his friend Dmitri Borgmann, a world expert on recreational linguistics, for supplying much of the information in his notes on word and letter oddities. "Jesus wept" (John 11:35) is identified as the Bible's shortest verse, and "Eber, Peleg, Reu" (1 Chron. 1:25) is cited as the shortest verse in the Old Testament. The Bible's longest verse is Esther 8:9. The longest word in the Bible, according to Dr. Matrix, is the 18-letter name Maher-shalal-hash-baz of Isaiah 8:1. The tallest man in the Bible is not Goliath (six cubits and a span, 1 Sam. 17:4), because we are told that Og, king of Bashan, slept in an iron bed that was nine cubits long (Deut. 3:11). The shortest man in the Bible is not Job's friend Bildad the Shuhite (Job 8:1), but Habakkuk, who wrote (Hab. 2:1), "I will stand upon my watch . . ." The smallest animals in the Bible are the widow's mite (Mark 12:42) and the "wicked flee" (Prov. 28:1).

Accepting the terminology of David L. Silverman, Dr. Matrix defines a homoliteral passage as one in which every pair of adjacent words has at least one letter in common. The longest such passage, he maintains, is Matthew

1:11–16, a fifty-eight-word sequence beginning "to Baby-
lon" and ending with "begat Joseph the husband." Dr.
Matrix attributes the 1973 discovery of this passage to Andrew Griscom of Menlo Park, California. When every adjacent pair of words has no letter in common, the sequence
is called heteroliteral. The longest passage of this kind
known was discovered in 1975 by Tom Pulliam of Somer-
set, New Jersey. It is a sequence of eighteen words from
Psalms 62:1–2, starting with "my salvation" and ending
with "I shall not be."

The practice of "consulting" the Bible for help in solving a personal problem was common throughout the Middle Ages. The ritual varied, but those who took it seriously
usually spent several days in prayer and fasting before
opening the Bible at random, then reading the first passage that caught their eye. Similar practices were common
in non-Christian cultures. The Greeks consulted Homer,
the Romans consulted Virgil, the Moors consulted the
Koran, and so on. Although from time to time the Church
issued decrees forbidding the practice, medieval history is
filled with stories about how lives were dramatically altered by it. Dr. Matrix identifies many biblical verses that
played such roles, notably Romans 13:13, 14, to which
Augustine attributes his conversion. As Augustine tells it
in his *Confessions,* book 8, he was sitting under a fig tree,
greatly agitated, when he heard a child's voice singing
"Tolle lege, tolle lege" ("Take up and read"). He picked
up a copy of Paul's epistles, opened it at random and read
the passage that transformed his life.

I had not been aware until I saw Dr. Matrix's comments
on Proverbs, chapter 11, that the chapter, with its thirty-
one verses, was widely used in medieval fortune-telling.
One simply consults the verse corresponding to the day of

the month on which one was born, then interprets the
verse as an omen. Dr. Matrix's example is the birth date of
Richard M. Nixon, January 9, for which the appropriate
verse is: "An hypocrite with *his* mouth destroyeth his
neighbour: but through knowledge shall the just be deliv-
ered." One could take this as an indictment of Nixon's role
in Watergate, but Dr. Matrix adds that according to a
friend of Nixon's the passage describes the persecution of
Nixon by his enemies and predicts the ex-president's
eventual vindication.

Hundreds of Dr. Matrix's notes give anagrams on biblical names and phrases. "Naomi" (Ruth 1), widowed and
bereft of her sons, becomes "I moan." For "ten commandments" the anagram is "Can't mend most men." For
"silver and gold" (Deut. 17:17) it is "grand old evils." For
"The wages of sin *is* death" (Rom. 6:23) it is "High fees
owed Satanist."

There are even notes on puns. "And he spake to his
sons, saying, Saddle me the ass. And they saddled *him*" (1
Kings 13:27). Dr. Matrix professes to find a defense of cigarette smoking in "Rebekah lifted up her eyes, and when
she saw Isaac, she lighted off the camel" (Gen. 24:64; see
Figure 31). We all know that when Jesus changed Simon's
name to Peter and said, "Upon this rock I will build my
church" (Matt. 16:18), he was aware that Peter meant rock
in Greek. But how many know that some of Jesus' sayings,
in the Aramaic dialect he spoke, contain witty wordplay?
Dr. Matrix points out that in Jesus' warning against the
hypocrisy of straining at a gnat and swallowing a camel
(Matt. 23:24), the Aramaic words for gnat and camel are respectively GALMA and GAMLA.

Occasionally, in commenting on the first mention of biblical personages, Dr. Matrix inserts old enigmas and re-

Figure 31.   Rebekah lighted off the camel (Engraving by Doré)

buses that have attained classic status. I shall cite only
three. Readers may enjoy identifying the three individ-
uals. (See Answers, Eighteen, IV.)

1. Five hundred begins it,
            Five hundred ends it,
        Five in the middle is seen.
    The first of all letters,
            the first of all numbers,
        Have taken their places between.
    And if you correctly this medley
            can spell,
    The name of an eminent king
            it will tell.

2. A Bible character without a name,
    Whose body never to corruption
            came,
    Who died a death that none
            had died before,
    Whose shroud is sold in every
            grocery store.

3.   H

19. Calcutta

"Magic makes you question the world. It raises your consciousness. I want to weave together a show so fantastic that people will question their reality." So the *New York Times* for July 9, 1976, quoted the magician Doug Henning. The new popularity of magic in the U.S. is surely a spin-off from the occult revolution and represents another response to the public's growing hunger for miracles and mystery. As the cults that feed these hungers continue to expand, many of them imported from the East, a colorful new consciousness-raising movement is spreading like a forest fire through the Bengal region of India.

The movement is called PM, an abbreviation for Pentagonal Meditation. Perhaps one reason it is so little known in the West is that it had its origin only a year ago in an obscure temple of Shiva on the outskirts of Calcutta, that vast megalopolis that most tourists in India avoid like the plague.

I first heard about PM from my old friend Sam Dalal, a Calcutta magician ("Sam the Sham") who edits (in English) a lively little magic periodical called *Mantra,* to which I subscribe and occasionally contribute. A newpaper clipping from Sam carried a picture of PM's founder, Guru Marahashish, standing in front of his temple with Zuleika, his assistant. It was hard to make out Marahashish's features because of his bushy white beard and mustache and the long white hair

190

that obscured the sides of his face, but Zuleika's smiling countenance was unobstructed. Although her skin was dark and the caste mark of Shiva—three horizontal lines—decorated her wide forehead, there was no mistaking those lovely Japanese eyes. It was Iva, the Eurasian daughter of Dr. Matrix!

How appropriate, I thought. Remove the first two letters of "Shiva" and you have Iva. I cabled Sam that I would be in Calcutta the following Monday.

It had been ten years since I had last visited Sam, but as the bus creaked and crepitated its way from Dum Dum Airport to the Grand Hotel the old familiar smell of Calcutta invaded the open windows. It was 4:00 A.M.—hot, foggy and windless. Except for a few new tall buildings the city looked the same. The pavements were littered with the pitiful bodies of the poor, lying under their dirty cotton cloths like corpses under shrouds. Indeed, by daylight many of the bodies would turn out to be corpses. A few cows wandered about looking for garbage. Here and there an early riser was washing himself at a fire hydrant or relieving himself in an alley.

Does any city in the world bring a visitor more starkly face to face with hunger, suffering, and death? In Calcutta there is only one way to avoid going mad. You must look on the city as you would a Hollywood set. None of it is real. It is a horror show in living color, thrown on a wide screen for your loathing and fascination.

I found Sam later that morning in his cluttered office on the second floor of a low building in the heart of the city. He too had not changed noticeably: he was still a slim, energetic young man with handsome features, a black mustache and goatee and coal black eyes. Sam is a Zoroastrian Parsee from Bombay. Like all magic buffs we wasted no time on idle conversation but immediately sat down to exchange the latest card tricks.

Sam was eager to meet Dr. Matrix. He knew that the

great numerologist had in his youth been an assistant to Tenkai, a renowned Japanese magician, and was said to be skilled in the conjuring art. After lunch we got into Sam's black Fiat and started for the temple, about an hour's drive.

Sam steered his way carefully through the choking traffic and hand-pulled rickshaws. We entered the Maidan, Calcutta's magnificent central park, headed north on Red Road and then swung west to cross the Howrah Bridge. Hundreds of Hindus were bathing below us in the Hooghly River. A bloated white cadaver bobbed downstream—one of the city's poor whose body had been tossed into the river like a piece of factory waste.

A few thousand Hindus die every year from drinking the pestilential holy water, but what does it matter? There is a popular theory among Calcutta doctors, particularly the homeopaths, that the people who survive gain immunity from Indian diseases.

At the newly renovated temple of Shiva a barefoot Indian boy in spotless white shorts stopped us at the entrance. I was an American journalist, Sam explained, who wanted to write about PM. Could I have an audience with Guru Marahashish?

"We are honored by sahib's visit," said the boy, bowing low.

He strode over to a large bronze *nataraja*, India's traditional statue of a dancing Shiva, to push a button at the center of the god's forehead. Chimes sounded in the temple. A moment later Iva herself, wearing a pentagonally tessellated sari, opened the door.

"Holy cow!" she exclaimed.

We embraced (Iva smelled much better than Calcutta), and after I had introduced her to Sam she led us to her father's office. The temple floor had been recently tiled in a curious periodic pattern (see Figure 32). The pattern seemed to depict a three-dimensional structure, but when you studied it carefully, you saw that the blocks were in an impossible arrangement. The structure could not exist. Huge mirrors covered the walls and ceilings. In

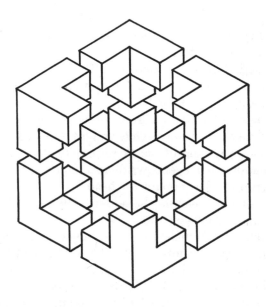

Figure 32. The basic region of Dr. Matrix's
impossible floor tiling.

every direction I could see hundreds of images of Sam, Iva,
and myself stretching off toward infinity.

At the entrance to a passageway Iva called our attention
to a wood statue of the four-armed Kali, the "bad" form of
Shiva's consort and the mother goddess of Calcutta. Legend
has it that Kali was once dismembered by Vishnu and one
of her fingers fell on the spot that became Kali-cutta: the
field of Kali. The goddess's skin was painted black. A scarlet
tongue hung from her grinning mouth. At each earlobe a
hanged man dangled, and her ebony neck was ringed with a
necklace of human skulls.

"Who's the prostrate person she's dancing on?" I asked.

"Her husband," said Iva. "Who else? I'm surprised *Ms.*
magazine has never put her on its cover. She's very popular

here."

"I sent you a copy of my new Scribner's book, *The Incredible Dr. Matrix.* Did you get it?"

Iva nodded as she opened the office door. "Then I lost it. Once I put it down I just couldn't pick it up again. But you boys must excuse me. I'll see you later."

Dr. Matrix's tall figure rose from behind his desk. His hair and beard were dyed snow white and his skin dark brown, but there was no way short of plastic surgery to alter his enormous hawklike nose. Black eyebrows divided emerald eyes from the painted stripes of Shiva on his forehead.

"Welcome to Calcutta," he said in a clipped British accent. "Sit down and I'll tell you about PM."

It was in the Himalayas, as a disciple of Swami Fonda-hondashankarbabasaranwrapi, that Dr. Matrix had learned the five basic principles of PM. The swami, whose smiling oil portrait was on the wall behind Dr. Matrix's desk, called his approach Basic Meditation. Dr. Matrix had changed the name because its abbreviation would give rise to gibes in English-speaking countries.

"The first principle of PM," said DM, "is: What is, is not. We call it *nonest.* The universe, including you and me, is no more than a monstrous stage illusion conjured up by Brahma while his colleague Shiva dances in the wings. It is all beautifully symbolized in the *nataraja* (See Figure 33). The universe has just begun one of its endless cycles. The ring of fire is the fireball of the 'big bang.' The drum in Shiva's upper right hand beats out the nonestic rhythms of space-time. The flame in his upper left hand is the energy that upholds and finally devours the world. His lower right hand is raised in a gesture meaning 'Fear not.' His lower left hand points to an uplifted foot that signifies release from the bondage of *maya:* the powerful magic spell that causes the uninitiated to believe the world is real."

Dr. Matrix had picked up a small ivory *nataraja* on his desk. "Who's the dwarf under Shiva's foot? I asked.

Figure 33. Shiva, Lord of the Dance, as sculptured in bronze in India
about A.D. 1000

"He's the demon of *avidya*, or ignorance."

"I know about *avidya*," said Sam. "Only Brahman-
Atman is real. The world doesn't exist except in the weak
way dreams exist. We are all phantoms in the mind of

Brahma, destined like the gods themselves to be absorbed back
into the One. Under the spell of *maya* we see the world
shattered into many parts. But the parts are only illusions
produced by the Great Magician."

Dr. Matrix nodded solemnly while he stroked his beard.
Then he clapped one hand and a young Hindu girl entered the
office bearing a glass goblet filled with pale red wine. Dr.
Matrix placed it in the center of a small circular table that
consisted only of a disklike top about a centimeter thick supported by a central rod about a centimeter in diameter and two
meters high. He covered the goblet with a tall blue cylinder
open at both ends.

"We'll come back to this in a few minutes," Dr. Matrix said.
"The second principle of PM is what we call the 'nonest giggle.'
It goes like this." Unsmiling, he gave out a high-pitched cack-
ling sound.

"What on earth does that mean?" asked Sam.

"That," replied Dr. Matrix, "is our technique for reacting to
anything that threatens to arouse our emotions. Since nothing
exists, there is nothing 'out there' to disturb us. If we pass a
crippled, blind or starving beggar, we giggle. If we pass a
corpse, we giggle. If we get a headache or any kind of pain,
the nonest giggle makes it disappear. Nothing exsits, therefore
the world is perfect. It is what it is not, so why try to change it?
As soon as a student of PM grasps this great truth we say that
he or she has 'lost it.' Only by losing all illusions can one find true
inner peace and improve one's tennis game."

"Your third principle?"

"It is the sacredness of five—the number of letters in Shiva.
Five is the middle digit in the sequence 1 through 9. It is the
central digit of the *Lo Shu*, the ancient Chinese magic square of
order 3."

"I remember," I said, "that back in 1966, before pi was
calculated in France to a million places, you predicted that the
millionth digit, counting the first 3 as a digit, would be 5." (See

my *New Mathematical Diversions,* page 100.)

"And I was exactly right, was I not? Five is a remarkable number. You must know that in the seventeenth century Sir Thomas Browne wrote an entire book about the ubiquity of fiveness." (I did not know, but I later checked it out. The book's full title is *The Garden of Cyrus or The Quincuncial Losenge or Network Plantations of the Ancients, Artificially, Naturally, Mystically Considered.*)

"Let's see," I mused. "There are five Platonic solids."

"Yes," said Dr. Matrix, fingering his beard. "And five points determine a conic curve. The fifth degree is the lowest-order equation that canot be solved in terms of radicals. The number of divisions necessary to find the greatest common divisor of two numbers is never larger than five times the number of digits in the smaller number. All groups of the order five or less are commutative. There are many other examples."

"Don't forget," said Sam, "that we have five fingers to a hand and five toes to a foot. A starfish usually has five arms. Most flowers have five petals."

"How does five enter into PM training?" I asked.

"It is what our neophytes meditate about. Five times a day, for a five-minute period, they assume the lotus position, close their eyes, breathe through their left nostril and say the world 'five' over and over again to themselves while they picture its numeral in their mind. At the end of each period they chant a secret mantra that we give them as soon as they have brought their teacher three ceremonial gifts."

"What kind of gifts?"

"One is a pocket mirror. Its mirror-image world signifies the illusory character of the world it reflects. Another is a banana. It symbolizes the *shivalinga,* the phallic symbol of Shiva that is worshiped throughout India. Sexual delight, you know, is one of Brahma's greatest illusions. The third gift is the equivalent of an American $50 bill. It is the first of five payments of $50 each. Teachers must donate a fifth of each payment to our temple."

"Only a fifth?" Sam asked.

"Yes. Zuleika giggles all the way to the bank. Our fourth principle is the doctrine of eternal recurrence. We teach that everyone has a thetan body (Christians call it a soul) that goes through five incarnations, each in a different cycle of the cosmos. After each universe is danced into nonest by Shiva it expands for 50 billion years. Then it contracts for 50 billion years and finally enters a black hole."

"The Black Hole of Calcutta?"

Dr. Matrix ignored my remark, but Sam winced. "After five cycles the universes repeat themselves. Your sixth incarnation puts your thetan back in the first universe, inside the same physical body you had before. We symbolize this with a curious numerical sequence I learned from my juggler friend Ron Graham when I visited him last year at Bell Laboratories in Murray Hill, New Jersey. Do you have a pocket calculator with you?"

"I'm never without one, or a deck of cards," I said, taking an inexpensive eight-digit calculator out of my shirt pocket.

Dr. Matrix handed me a pad and asked me to jot down any two real positive numbers. I wrote pi and 76. He asked me to add 1 to the second number and then divide the sum by the first number to get the third number of the sequence. I entered 76, added 1 and divided 77 by 3.1415926. The result: 24.509861. To get the fourth number I repeated the recursive procedure: add 1, then divide by the previous number. This yielded .335656. The fifth number was .0544946.

"Now," said Dr. Matrix, "for the zonk. You would probably guess that if you continue applying the algorithm, you will get more and more horrible numbers. But try it once more and see."

I was zonked all right. The next number was pi! The readout was 3.1415931, but that was because tiny errors had accumulated in the machine. Dr. Matrix assured me that when all calculations are accurately made, the sequence loops with a period of five. It is easy to prove, but I leave the proof as a problem for the reader.

Dr. Matrix showed us several other calculator curiosities involving five, but I have space for only one more. Enter 555 and divide five times by 5. The results are 111, 22.2, 4.44, .888, and finally the year of the American independence, .1776. Try starting with a row of as many 5's as your readout takes.

"And the fifth principle?" asked Sam, who had been taking notes.

"It is our Supreme Principle. We reveal it only after a student has made his fifth payment."

"What's happening there?" I asked, pointing to the blue cylinder.

"Something nonestic," answered Dr. Matrix. "Kindly lift the tube."

When I did, I could not believe my eyes. The goblet and its contents had vanished! I examined the cylinder. There was nothing suspicious. Sam was smiling enigmatically. Later he told me that he himself had recently invented this trick. Dr. Matrix must have learned it from a local conjuror. Can the reader guess the clever modus operandi?

"The goblet and the wine never existed," Dr. Matrix said. "They were just illusions." He giggled nonestically. "But let me show you something you'll like even better: one of the ancient dances of Kali."

He led us out of his office and down a corridor to a small theater-in-the-round. We seated ourselves in the lowest circle. Dr. Matrix again clapped one hand (we never did figure out how he made the sound), and the bright light was instantly replaced by a dim red glow from concealed bulbs at the ceiling's circumference.

From somewhere came the twangy sound of an Indian *raga* played by native instruments. The beat seemed to get constantly faster, but this was an illusion because it always remained the same. Suddenly Iva appeared on the small stage. Her only costume was a pentagonal mirror on her navel. It glinted and flashed ruby light while her torso swayed back and forth in a sensuous dance of indescribable beauty. In the scarlet light her skin was as black as the skin of Kali.

"Zuleika is topless and bottomless," said Dr. Matrix, "to remind us that we must strip ourselves of all illusions. The world has neither top nor bottom because it doesn't exist."

Sam and I applauded widly when the dance ended. "Viva illusion!" I shouted as I stepped forward to give Iva a congratulatory hug. My arms passed through thin air. Her image had vanished, although I could hear her giggling nonestically in the distance. It was all a stage trick produced by hidden concave mirrors.

Iva returned to the city with Sam and me, and the three of us had dinner at a Japanese restaurant near the Grand. "Let us eat, drink and be merry," said Sam as we raised our cocktails, "because nothing is real."

"And because," added Iva, "tomorrow I may diet."

She told us that her father was organizing a major PR campaign to promote PM in the U.S. The campaign will be managed by Bagel Lox, a former vice-president of the Dr. Pepper Company. Jerry Rubin, Mrs. and Mr. John Lennon, Doug Henning, and Mia Farrow have already visited the Calcutta temple (all expenses paid) and have become enthusiastic converts. John Denver is writing a song about PM called "Don't Worry about What Ain't."

I would have stayed longer in Calcutta if Iva and her father had not planned to leave in a few days for a visit on L. Ron Hubbard's yacht. It had just entered the Bay of Bengal to tie up at what Kipling called "The City of Dreadful Night." Before I left the sad yet somehow beautiful city Iva handed me a folded file card on which she had written my secret mantra.

As my plane climbed above the Nonestic Ocean I ordered a martini and glanced once more at my mantra. *Ohwa—taboo—*

biam. My consciousness was being altered by a second cocktail before the meaning hit me. I fancied I could hear Iva giggling, but it was just an auditory illusion.

20. Stanford

> The emotion consisted wholly of glee and admiration; glee at the vividness which such an abstract idea or verbal term as 'earthquake' could put on when translated into sensible reality and verified concretely; and admiration at the way in which the frail little wooden house could hold itself together in spite of such a shaking. I felt no trace whatever of fear; it was pure delight and welcome.
>
> '*Go it*,' I almost cried aloud, 'and go it *stronger!*'
>
> —William James at Stanford University, 5:30 A.M., April 18, 1906, reacting to the San Francisco earthquake

"In vulgar language," writes Garrett Hardin, professor of human ecology at the University of California at Santa Barbara, "we need an earthquake-predicting facility like we need a hole in the head."

This quotation is from Hardin's essay "Earthquakes: Prediction More Devastating than Events," which is reprinted in his splendid collection of papers, *Stalking the Wild Taboo* (William Kaufmann, 1973). Hardin believes that if geophysicists ever succeed in predicting earthquakes with high accuracy, the social disruptions produced will do more damage than the quakes.

By a curious synchronicity I had just finished reading Hardin's essay when a letter arrived from an old friend and magic enthusiast, Persi Diaconis, now a statistician at Stanford University, Persi wrote to tell me that a man by

202

the name of Dr. Punk Rockwell had rented a small office building on El Camino Real, between Redwood City and Menlo Park. The building was about 25 miles south of the heart of San Francisco and only a few minutes drive from Stanford. Dr. Rockwell was head of a firm called the Punk Earthquake Prediction Corporation (P.E.P.C.), which claimed to have a new, infallible method for predicting quakes.

For the past several months, Persi went on, Rockwell and his lady assistant, Punky Anderson, had been selling their service to farmers and other residents all along the California coast, from Eureka to San Diego. It was Dr. Rockwell's contention that some of the mechanical stresses that have been building up for decades along the 600-mile San Andreas fault—in particular those resulting in the recent upward bulging of the area surrounding Palmdale—would be suddenly released late in December 1977. To anyone willing to pay a modest fee of $1,000 Dr. Rockwell would disclose the exact time and severity of the coming quake at any designated spot within 20 miles of the fault.

Persi's brief description of the Rockwell prediction method was so outlandish that I assumed he was putting me on. Then came the final revelation that made it all believable. "Dr. Rockwell," Persi added, "is a tall, elderly gentleman with a large hawklike nose and glittering green eyes. His assistant has unmistakably Oriental features. The pair are obviously none other than the notorious numerologist Dr. Irving Joshua Matrix and his Eurasian daughter Iva."

Whenever Dr. Matrix and Iva surface with a new swindle, I drop everything to pay them a visit. I have learned never to alert the pair to my arrival ahead of time. Although they have found me trustworthy in the past, how can they be certain I will not reveal their identity to the local police? For this reason I have found it advisable to take them by surprise.

At the San Francisco International Airport I rented a car and drove south along the Bayshore Freeway. Persi had offered to put me up for a few days at his house near Stanford. It was a fine, chilly November night, a pale fingernail moon floating low over the dark ominous waters of San Francisco Bay.

Persi and I stayed up late going over new variations of the Zarrow shuffle and the Ascanio spread and discussing Persi's unpublished work on the probabilities of ESP card testing. The next morning we drove to the nearby P.E.P.C. headquarters. Pressing the entrance bell produced the sound of muffled drumbeats behind the door.

Iva, as usual, looked startled when she opened the door. Then she flashed one of her seductive Oriental smiles and we exchanged kisses on the cheek before I introduced her to Persi.

I must have looked even more surprised than Iva. She was wearing the dirtiest pair of blue jeans I have ever seen. The fly was half-open, fastened with a large safety pin. A long tear in one jean leg disclosd 10 inches of bare thigh. Her shredded yellow shirt was dotted with black swastikas and the shirttails were tied in a loose knot above her bare midriff. Her hair, dyed a pale green, fell unkempt around her shoulders. A gold safety pin swung from each ear lobe. I had read about the new "punk look" that was spreading from London to California's subculture, but it was even punker than I had imagined.

"Dr. Rockwell is attending a conference at the Stanford Research Institute," Iva said, "but he should be back within the hour." Several seismologists from the Earthquake Research Center at Menlo Park and a top official of the Department of Defense were attending the conference. There was a good possibility, she hinted, that the SRI might obtain substantial government funding for research on Dr. Rockwell's techniques.

Iva led the way to a small laboratory. Its walls were lined with hundreds of glass tanks containing what I guesed to be at least a million cockroaches. Persi and I listened with amused smiles while Iva solemnly explained how the Rockwell system operates. It was, she said, Dr. Helmut Schmidt who had made the initial breakthrough.

I nodded. I was quite familiar with Schmidt's work. He is a
Ph.D. physicist who used to be the director of Dr. J. B. Rhine's
parapsychology laboratory in Durham, N.C. I remembered that
about seven years ago Schmidt had reported that cockroaches
seemed to have the psychokinetic (PK) ability to cause a random-
izer to give them electric shocks more often than chance allows.
Schmidt admits, however, that he hates cockroaches. Since cockroaches presumably do not enjoy being shocked, Schmidt and
his colleagues suspect that it was he who was the PK agent. The
accepted theory is that cockroaches probably have some PK
ability, but it is overwhelmed by the stronger, unconscious PK
influence of experimenters who are not fond of cockroaches.

Iva explained that Dr. Rockwell was the first to prove that in
addition to their PK ability cockroaches have powers of precognition. His now classic experiment utilized a large box with
doors at opposite sides, each leading to food. At one-hour intervals one of the two doors was opened at random, the choice
being made by a computer using a tape of prerecorded random
binary digits. Prerecording the digits, Iva added, eliminated the
possiblity that the cockroaches were employing PK to influence
the randomizing process. The insects quickly learned to anticipate which door would open next. For several minutes before a
door opened large numbers of roaches would swarm in front
of it.

"How can you be sure," Persi asked, "that the roaches aren't
picking up numbers on the tape by clairvoyance?"

"Good question," replied Iva, with a dim smile. Dr. Rockwell's later earthquake-prediction experiments, she explained,
ruled out both clairvoyance and PK. "After all," she said, "it
would be ridiculous to suppose that cockroaches have enough
PK power to cause an earthquake."

"I wouldn't rule it out," I said. "Don't forget that severe
quakes reduce supermarkets and kitchens to shambles and provide marvelous new food supplies. Millions of roaches, acting in
telepathic union, might work up enough reinforced PK power

to trigger action along a fault. Only a tiny kick is needed, you know, to snap the tension."

Iva gave me a cold stare. "*You* need a big kick," she said. She went on to explain Dr. Rockwell's sensational discovery of last January. Following the lead of a Russian parapsychologist who had been doing secret work on insect precognition, Dr. Rockwell found that the precognitive power of cockroaches can be greatly enhanced by two things: large doses of vitamin B-1 and smoke from burning punk.

Fortified by B-1 and drunk on punk, cockroaches become extraordinarily sensitive to future earth tremors. A careful monitoring of the animals' agitation by ingenious optical instruments provides accurate measurements of both the time and the intensity of a future quake. The length of time until the next quake is a function of the average distance a roach travels before it turns. The severity of the quake is a function of the average speed of the roach.

Iva told us that Dr. Wilhelm J. Levity, an American parapsychologist who recently emigrated to Romania, had been the most successful in replicating Dr. Rockwell's B-1-punk experiments. Working with Romanian roaches, Dr. Levity had predicted two weeks in advance the day and hour of the Romanian earthquake of March 1977 that almost destroyed central Bucharest and killed hundreds of people.

"Wouldn't that make him a Romanian hero?" Persi asked. "How come we haven't heard of him?"

"The Romanian government has clamped tight security on Levity's work," Iva answered. "We've been trying to reach him. At the moment we don't even know if he's alive or dead."

"What do your roaches tell you about the coming California quake?" I asked.

"We know the exact day and hour," Iva said, "but we divulge this information only to our paying customers. I can tell you, though, that it will be before the end of December, and that horizontal slippage along parts of the San Andreas fault will be as much as six feet. But here comes Dr. Rockwell."

Brisk footsteps sounded outside the lab, and in strode Dr. Matrix. He was wearing tight black pants that appeared to be made of plastic garbage-bag material. A dirty T-shirt bore the smiling face of Dracula. Dr. Matrix's hair had been dyed a bright pink. A large Band-Aid on his forehead seemed to cover a fresh wound.

"So it's you again," Dr. Matrix said, his emerald eyes flashing mild hostility above a pair of half glasses. He turned toward Persi. "Dr. Diaconis, I presume."

"How did you know?" said Persi.

"I heard you lecture at Stanford last week on the probabilities of poker. And I have a friend—a dealer at Lake Tahoe—who says you do the second-best second deal west of the Mississippi."

"And who does the best?" said Persi, smiling.

"I do," said Dr. Matrix.

Iva excused herself. We followed Dr. Matrix to his office at the rear of the building, where we sat and chatted for more than an hour. A Confederate flag stood at one side of his desk. On the back wall a large portrait of President Millard Fillmore beamed down on us.

"You may wonder," said Dr. Matrix, "what connection there is between our punk attire and earthquakes. It's more than just the use of punk to raise the psi level of cockroach consciousness. The punk movement is a refreshing protest against life's terrible unfairness. It's a punk, punk, punk, punk world. Can you think of a better symbol of nature's cruelty than a monstrous earthquake that suddenly snuffs out tens of thousands of lives? Unfortunately next month's California quake will be relatively mild."

"According to *The Jupiter Effect*, by John Gribbin and Stephen Plagemann," I said, "Los Angeles will be totally destroyed by a quake in 1982."

"Sheer balderdash," snorted Dr. Matrix. "Gribbin and Plagemann think the quake will be triggered indirectly by Jupiter's gravity when it is reinforced by the gravity of all the other planets lined up on the same side of the sun in 1982. Jupiter's gravity is supposed to alter the sunspots, which in turn are supposed to alter the composition of the earth's atmosphere and produce the quake. Did you notice that Gribbin and Plagemann's book has no diagram showing how the planets will line up? That's because they won't line up. They will get to the same side of the sun—Gribbin and Plagemann are right about that—but they'll be so dispersed from a straight line that a picture would demolish his thesis. Besides, there's little evidence that the planets can affect sunspots, and no evidence sunspots can affect the earth's atmosphere enough to disturb the San Andreas fault. Gribbin and Plagemann's book is garbage. According to our cockroaches, Los Angeles won't be destroyed until a major quake hits it in the prime year 1987."

The number 1,987, Dr. Matrix noted, is a remarkable prime. If we look for primes with positive digits in a cyclic, descending, consecutive order, 1,987 is the lowest example after 19 and 43, and the highest before 76,543, the largest-known prime of this type. If 0 is allowed, there is 109 and 10,987. Primes with digits in cyclic ascending order are commoner. Nineteen are known, starting with 23, including 23,456,789 and 1,234,567,891, and ending with the astonishing 1,234,567,891,234,567,891,234,-567,891. The 28-digit prime was discovered in 1972 by Raphael Finkelstein and Judy Leybourn of Bowling Green University. Can the reader prove there are no descending consecutive-digit primes, with or without 0, that start with 9?

Dr. Matrix reached into a desk drawer, pulled out a paper-covered booklet and tossed it across the desk. Persi and I twisted our necks to read the title: *Alphabetic Number Tables, 0-1000*. The preface begins: "It gives us great pleasure, not unmixed with profound emotion, to at last make public these alphabetical tabulations of the natural integers, ordered both in English literation and in Roman numerals." The booklet had been published at the Massachusetts Institute of Technology on April 1, 1972.

"I compiled both lists for some friends at M.I.T.," Dr. Matrix said. "You may keep the monograph if you like. It's now quite rare. I give it to you because it suggests some interesting problems."

I opened my notebook again and raised a pencil.

"The alphabetical list of the English spellings for the integers 0 throught 1,000 begins eight, eight hundred, eight hundred eight, eight hundred eighteen, eight hundred eighty and so on. The last entry, of course, is zero. How many of your readers can name the 1,000th, or next to last, number on the list?"

"Delightful," I said, scribbling a note.

With the back of his hand Dr. Matrix flicked away a large cockroach that was crawling across his desk. "You can ask a similar question for the Roman numerals," he continued. "The digits themselves are now the letters. The sequence begins: C, CC, CCC, CCCI, CCCII, and so on. The Romans had no zero, and so the last number is the 1,000th one. Can your readers determine what it is?"

Dr. Matrix paused until I had finished writing, and then went on: "The English spellings of numbers suggest strange problems. For example, consider the first letters of the names of the 10 digits. What is the longest English word that can be made with those letters? You needn't use all of them, and any letter may, if you like, be used more than once. I think the longest such word is 'festoons.' Perhaps your readers can find a longer one. Another interesting question. What's the smallest positive integer whose English name contains all five vowels plus y? Assume, as I did when I compiled the M.I.T. list, that 'and' is not a proper part of any number name."

"Excellent," I said. "Any others?"

"Numerological problems are as infinite as the primes," said Dr. Matrix. "But I'll give only one more. It's the invention of my friend Joe Wagner of New Wilmington, Pennsylvania."

Dr. Matrix picked up my notebook to jot down the following sequence: 10^3, 10^9, 10^{27}, 10^2, 10^0, ...

The problem, he explained, is to determine the exponent of 10 that gives the next number. The spellings of the numbers determine the pattern.

"I've been fascinated by those three cubes on your desk," Persi said, pointing to three large cubes on a tray at the top of a desk calendar. Each face of a cube bore a lowercase letter. The cubes were in a row so that the side facing Dr. Matrix spelled "nov."

"I remember," said Persi, "that in one of his columns Martin asked how to put digits on the 12 faces of 2 cubes so that they could be placed side by side to give any day of the month [see my *Mathematical Circus*, page 186]. Is it really possible to turn and rearrange these three cubes to spell the first three letters of any month?"

Dr. Matrix replied by extending 10 bony fingers and rapidly manipulating the cubes so that they ran through the standard abbreviations for all 12 months. It is a pretty problem to figure out how the lowercase letters can be distributed on three cubes, one letter to a face, to make this possible. I later learned that Dr. Matrix's friend W. Bol of Geldrop in the Netherlands had solved the problem and sent the cubes to him in 1971.

Dr. Matrix consulted a Richard Nixon watch that dangled at the end of a long dog chain pinned to a shoulder of his T-shirt. As it ticked, Nixon's eyes darted from side to side. "It's almost one-thirty," he said, pressing a button on the side of his desk. The air was instantly shattered by the voice of Johnny Rotten, the Irish punk rock celebrity, snarling out the lyrics of his latest British hit, backed by the stabbing, quaking three-chord beat of his band, the Sex Pistols. Iva materialized at the door.

"Where shall we all lunch, my dear Punky?" Dr. Matrix asked.

"There's a new place on the Camino," Iva replied, "called the Punk Elephant. The food is great, and they serve nonpunk cocktails."

"Will I have to listen to punk rock?" Persi asked apprehensively, cupping his hands over his ears.

"No," said Iva. "There's a punk girl group appearing there called the Palmdale Bulges, but they don't start until nine in the evening."

Iva vanished for a moment, then returned with two long black leather overcoats. She handed one to Dr. Matrix, who had walked over to stand facing her. And then we saw an astonishing thing.

Each of them held an overcoat in front of the other, then each simultaneously let go of the coat with his (her) left hand and pushed his (her) left arm through the left sleeve of the coat opposite. Still facing each other and acting in unison, each of them then put his (her) left arm around the other's right side and with his (her) left hand grasped the top of the coat as the right hand carried it behind the other's left shoulder. Each now dropped the coat from the right hand, carried the coat around the other's back with his (her) left hand, then each pushed his (her) right arm into the right sleeve. Now they were wearing their own overcoats, and they were still facing each other. The ritual was executed so rapidly that it was over before Persi and I could observe just how they did it. Iva told me later, when she helped me to write the above instructions, that it was an old Japanese vaudeville routine.

As we walked past the entrance to the laboratory I sniffed the air. "I smell punk," I said.

"Shhh," said Iva. "If you don't talk about it, maybe no one will notice."

21. Chautauqua

Computers don't actually think. You
just think they think. (We think.)
—Theodor H. Nelson,
Dream Machines

For years computer scientists who are also experts in linguistics have been working on systems to enable computers to converse with human operators in a natural spoken language. Progress on such systems has been disappointingly slow. As of today only the most trivial and stylized dialogue can be carried on, usually in the form of typed sentences with an extremely limited vocabulary.

On the other hand, science-fiction fans have been familiar with talking computers for more than half a century. The marvelous machines have even appeared in children's fantasy literature: as early as 1907 L. Frank Baum introduced (in *Ozma of Oz*) a mechanical windup robot named Tik-Tok, which according to its manufacturers could think, speak, act and do "everything but live." In recent years, through films such as *Forbidden Planet* and *Star Wars*, the general public has become accustomed to talking robots and to talking computers such as HAL, which ruled the spaceship in Arthur C. Clarke's *2001: A Space Odyssey.*

There are many signs of the new familiarity with such ideas. For example, consider a toy robot that recently went on sale: a child answers multiple-choice questions by pressing buttons on the robot's torso, and the robot comments (by a prerecorded tape) on the quality of the answers. Moreover, game-playing computer programs are improving rapidly. For less than $300 one can buy a small chess-playing computer that will defeat good players. More sophisticated chess programs are now approaching the master level. When the time limit on moves is short enough, they can trounce even a grandmaster.

For these reasons and others there was little skepticism when advertisements began to appear announcing public demonstrations of ASMOF, the world's first talking computer. The name ASMOF was an acronym for the American Superior Mind Operating Foundation, which was sponsoring demonstrations of the robot prototype to stimulate sales of stock. From newspaper clippings readers began to send me, I gathered that the robot featured a new magnetic-bubble memory. Its circuitry was inside a 20-foot-high aluminum figure designed to resemble a movie robot. The monster had two openings where a person's eyes would be and a third eye with a ruby-colored lens at the center of its massive forehead. There was no nose, but a conical loudspeaker formed a kind of mouth. At the public demonstrations the robot was seated behind a large table and did not move. Occasionally a beam of crimson light shot from the third eye to scan the table or anyone seated in a chair on the opposite side.

ASMOF began its tour of the country at the end of July, making two-hour appearances at halls and theaters in large cities and summer resorts. For an admission fee of $3, people in the audience were allowed to sit across from ASMOF at the table and ask it a reasonably brief question on any topic. The robot replied in a husky mechanical voice and sometimes engaged in humorous banter with a questioner before trying to answer the question. It did not know everything. Occasionally it responded with remarks such as "Your problem would take too long to compute," or "Sorry, madam, that information is not in my memory bank."

On special occasions an entire appearance was devoted to combat with a local expert in some intellectual table game. ASMOF seemed to be a top-level player of chess, checkers, go, and even games with random elements such as backgammon, bridge, and poker. Early in August, when ASMOF was at a theater in Nyack, N.Y., a chess grandmaster who lives nearby was offered $1,000 if he could defeat the robot. The grandmaster lost in 18 moves. ASMOF then challenged him to a second game, offering to play without its queen's bishop if this time the grandmaster would wager $1,000 on the game. The bet was declined.

The grandmaster's humiliating defeat was a startling turn of affairs. When I telephoned an old friend who works in the artificial-intelligence laboratory at the Massachusetts Institute of Technology, however, he assured me that ASMOF was a total fraud. He did not know precisely how the robot was controlled, but he was sure the foundation backing it was a swindle. He intimated that my old acquaintance, the numerologist Irving Joshua Matrix, might be behind the scheme.

Glancing over my supply of press clippings, I could find no pictures of the foundation's head, Frank Rossum, or his assistant, Josie Clarke Nelson. Knowing Dr. Matrix's obsession with wordplay, I began to think about the two names. Could Frank be an abbreviation for Frankenstein? Rossum certainly suggests Rossum's Universal Robots, from Karel Capek's 1920 play R.U.R. (The play introduced into the English language the word "robot," from the Czech robota, meaning work or compulsory service.) What about Josie Clarke Nelson? Josie might be a feminine form of Joe, Lewis Padgett's famous science-fiction robot, which could "varish" as well as "skren." Clarke, of course, would be for Arthur C. Clarke. And Nelson could refer to Theodor H. Nelson, the young computer scientist whose double book *Dream Machines* and *Computer Lib* is such a whopping success as a wild, Oz-like introduction to the fantastic new world computers are creating. (His later paperback *The Home Computer Revolution* contains still more startling predictions.)

But wait! There was another extraordinary coincidence. It was Dr. Matrix himself who first revealed that HAL is obtained by shifting each letter of IBM back one letter in the alphabet. Now shift each letter of IBM forward one letter. The result is JCN, the initials of Rossum's assistant!

From literature supplied by Rossum's foundation I learned that ASMOF'S next public appearance would be in the amphitheater of the venerable Chautauqua Institution on the shore of beautiful Chautauqua Lake in upstate New York. From my house in Westchester I drove there easily in a day, putting up at a motel close to the Chautauqua Institution's entrance. The next afternoon I made a point of arriving at the amphitheater early enough to be sure of a front seat. On the chance that Dr. Matrix or his daughter Iva might appear on the platform, I disguised myself with dark glasses and a large artificial moustache.

There was no sign of either of them. A pleasant young man with long black hair and a short beard introduced himself as a "futurologist" formerly with the Stanford Research Institute. He spoke learnedly about recent progress in computer communication in natural languages, and finally about the monumental breakthrough by computer scientists working for the American Superior Mind Operating Foundation. Their discoveries, which had been made possible by the new bubble memory, he explained, were still highly classified. He hinted darkly that the CIA was trying to prevent the manufacture of smaller versions of ASMOF. The robot, he assured us, unlike human beings, cats, and cockroaches, possessed no paranormal powers. Hence ASMOF would not answer questions about the future or questions that required any kind of extrasensory perception. The foundation was working closely with several physicists at the Stanford Research Institute on ways to add ESP and precognition to ASMOF'S circuitry, he added, but that development might take another decade.

It was a singular and entertaining performance. Most people asked trivial questions, such as what years Harding

was president or who won the baseball world series in such-
and-such a season—facts that could easily be stored in a
computer's memory. Some of the questions, however, were
harder. I took extensive notes on those involving the kind of
mathematics and wordplay I thought would most interest my
readers.

For example, a young woman asked ASMOF for the longest word in the English language. After scanning the woman
with its ruby eye and complimenting her on her smile,
ASMOF asked: "Can the word be hyphenated?" The woman
replied yes. "In that case," said ASMOF, there *is* no longest
word. We can speak of a great-grandmother, a great-great-
grandmother, a great-great-great-grandmother, and so on.
Next question, please."

Here are a few other linguistic queries of an unusual nature. A woman asked for a ten-letter word of one syllable.
ASMOF came up with "scraunched." A young man asked
for an English word containing the sequence of adjacent
letters "nkst." After some conversation about the slogan on the
questioner's T-shirt, ASMOF gave an acceptable reply. The
man was followed by his brother, who asked a similar question about the sequence "nksh." This too was correctly answered. Can the reader supply a dictionary word containing
each sequence?

Children liked to test the robot with ridiculous riddles.
Sometimes ASMOF answered correctly. When it failed, it always asked for the answer to store in its memory and occasionally praised the child for having stumped it. I find in my
notes that one elderly man, a professor emeritus of English at
the nearby State University of New York at Buffalo, asked a
literary riddle that was new to me. He wanted to know what
Coleridge's ancient mariner and an inebriated shortstop had
in common. ASMOF promptly responded: "He stoppeth one
of three."

The most difficult word question was asked by a Shakespearean scholar from Bryn Mawr. There is a line in one of Shakespeare's plays, she said, that begins with "My." Reading upward, the first letters of the preceding four lines spell WANT. Reading downward, the first letters of the succeeding four lines spell BABY, yielding WANT MY BABY. In what play does this acrostic appear?

This question must have been hard, because ASMOF spoke for several minutes about whether or not such Shakespearean acrostics were intentional. Finally the robot identified the line. It is the 14th line from the end of the first scene of the first act of *The Comedy of Errors:* "My soul should sue as advocate for thee." Is it possible, ASMOF asked, that Shakespeare was sending a coded plea to a lady friend, asking for possession of an illegitimate child?

Some questions involved formal logic. Ken Knowlton of Bell Labs tried to throw the robot's circuitry into an endless loop of alternating yeses and noes by posing Bertrand Russell's famous paradox about the barber in a certain town who shaves all the men and only those men in town who do not shave themselves. "Is the barber a self-shaver?" he asked. ASMOF replied: "Insufficient data. She could be."

What is the only way, a California mathematician named David Silverman wanted to know, to divide all positive integers into two disjoint sets in such a way that no pair of numbers in either set add up to a prime? The robot failed to solve this problem.

Thomas Szirtes of Montreal asked ASMOF if it were possible for him (Szirtes) to be exactly one-third Scottish, one-third Chinese, and one-third Hungarian. ASMOF replied that it was not. One has, explained the robot, 2^1 parents, 2^2 grandparents, 2^3 great-grandparents, and so on. Hence the question is equivalent to asking whether 2^n can equal $3x$. Now, 2^n equals $2 \times 2 \times 2 \times 2 \times \ldots$, with 2 repeated n times. Since 2 is a prime, we know from the fundamental theorem of arithmetic that this sequence is a unique factorization of

$2n$ into primes, that is, that $2n$ has no prime factors other than 2. Therefore 3 cannot divide $2n$, and the original conjecture must be false.

Jaime Poniachik of Buenos Aires, who edits an excellent Spanish puzzle magazine titled the *Snark*, happened to be visiting the Chautauqua Institution at the time of the ASMOF demonstration. When it was his turn to ask a question, he said he had a friend in the U.S. with a curious social security number. Its nine digits include every digit from 1 through 9. They form a number in which the first two digits (reading from left to right) make a number divisible by 2, the first three digits make a number divisible by 3, the first four digits make a number divisible by 4, and so on until the entire number is divisible by 9. What is the number? ASMOF expressed admiration for this new problem but pleaded insufficient time to compute the answer.

When it was my turn, I asked permisison to give a problem involving six chess knights on a three-by-four board. On a sheet of paper I sketched the initial position shown in Figure 34. The task is to exchange the white and black knights with a minimum number of knight moves. The knights may be moved in any order, regardless of their color, but no two of them may occupy the same cell.

In 1974, when this problem was published in *Journal of Recreational Mathematics*, it was called trivial and a 26-move solution was given. Later an 18-move solution was found. I included the problem in the first edition of *Aha!* (the handbook for a set of six high school filmstrips called *The Aha! Box*) and cited the 18-move solution as the best. Three readers of the book, Gary Goodman, Warren B. Porter and George Schneller, independently reduced the solution to 16 moves!

After scanning my sketch with its third eye, ASMOF offered the 18-move solution and expressed doubt when I said there was a shorter one. When I gave the shorter

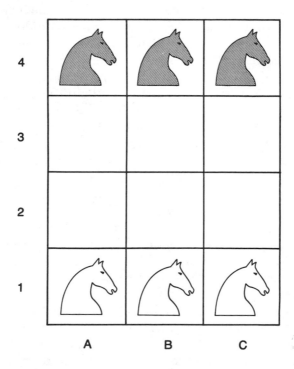

Figure 34. A knight-switching problem

solution ASMOF congratulated the discoverers for their "Aha!" insight concerning what seems to be a wasted move. Can readers find the 16-move answer?

After watching ASMOF perform for about an hour I was convinced that my friend at M.I.T. was right. None of the calculations were being made by internal circuitry. The robot was surely under the control of a human intelligence at some nearby spot. I had learned from an ABC television crew, there to film the performance, that Rossum and Nelson were staying at a hotel on the Chautauqua Institution grounds. I slipped out of my seat and left the amphitheater. When I arrived at the hotel, the lobby and corridors were deserted. Everyone, it seemed, had gone to the demonstration. No one saw me put my ear to the door of Rossum's room. Inside I could hear the faint sounds of conversation.

The hotel is an old one, and I had little difficulty snaking a plastic credit card through the crack of the door and forcing back its spring catch. The scene inside the room was not very different from what I had expected. Iva was seated at a large desk with enormous headphones over her ears. A color television screen in front of her displayed the questioner sitting opposite ASMOF in the amphitheater. I later learned that the television camera was behind the robot's left eye. On both sides of Iva's desk were the same basic reference books that librarians keep on hand for answering telephone queries: *The World Almanac, The New Columbia Encyclopedia,* several *Who's Whos,* an unabridged dictionary, concordances of the Bible and Shakespeare, dictionaries of quotations, and so on. Also close at hand were the console of a large programmable computer and a telephone into which Iva could insert cards that automatically called numbers.

Iva looked startled when I entered. She took off her earphones and put a hand over the microphone in front of her. "If you don't leave at once," she said angrily, her dark eyes flashing, "I'll scream and call the police."

"I think not," I said, removing my dark glasses and ripping off my moustache.

"So, it's you again!" said Iva with a twisted smile. "Sit down and keep your big mouth shut."

I sat on the edge of the bed and watched her operate for the rest of the hour. An electronic device completely disguised Iva's pleasant voice, giving it that cold, toneless timbre characteristic of television and movie robots. Easy questions from simple souls were answered from the considerable fund of information inside her own head. Other answers she looked up quickly in the reference books on the desk while she killed time by engaging in conversation with the questioner. For problems requiring numerical calculations, such as finding the 13th root of 10 or raising pi to the power of pi, she turned to the computer. When a more difficult question was asked, she inserted a card into the telephone. By restating the problem aloud—ostensibly to make sure it was correctly understood—she would ask it of the person she had called. Iva later told me that Dr. Matrix had hired more than 100 experts, each one a specialist in a different field, and was paying them handsomely to be on hand at private telephone numbers during the demonstrations.

Iva also told me that Dr. Matrix had tried to hire Isaac Asimov for her job to avoid having to pay so many experts. Asimov had politely declined, however, mainly because he had to complete seventeen books before the end of the year. I never learned the name of the mathematics expert. Nor was I able to discover which grandmaster had been hired to play the robot's chess games.

When the two-hour performance ended, Iva removed her earphones, switched off the mike, and grinned. She squinted at her wristwatch, one of those tiny watches that are so expensive the dial is almost impossible to read without a magnifying glass. "Time for cocktails," she said.

Dr. Matrix, I was delighted to learn, had left for the day. The old buzzard was visiting a friend who worked as a phony medium in the nearby Spiritualist community of Lily Dale. No restaurant on the Chautauqua Institution grounds is allowed to serve liquor (a holdover from the institution's pious past), and so Iva guided us northward on Highway 17 to a colorful seafood restaurant in Westfield, on the shore of Lake Erie. (I was amused by a statement on their menu: "We serve oysters in January, February, March, April, Mayr, Jurne, Jurly, Augurst, September, October, November, and December.") We ordered dinner.

"In my opinion," I said between mouthfuls, "this is the salad of a bad café."

Iva winced at the Spoonerism. "I think the salad's excellent."

"Well," said I, "there are two sides to every question."

"*Au contraire*," she repiled. "There are *not* two sides to every question."

Nevertheless, we had an excellent dinner and an enjoyable evening. A month later, when ASMOF was booked into a theater in Washington, D.C., two investigative reporters from the *Washington Post*, aided by Randi the magician, figured out the con. They rented a hotel room next to the one from which Iva was operating, bored a hole through the wall, and obtained good photographs of her at the microphone. The story broke in the next morning's newspaper, but by then Dr. Matrix and Iva had disappeared. ASMOF was left behind in the van they had used for transporting it. The Washington police have since given the robot to the Smithsonian Institution, where it will soon be on display with a tape-recorded account of the swindle.

22. Istanbul

> The sons of the prophet are brave men
> and bold,
> And quite unaccustomed to fear,
> But the bravest by far in the ranks of
> the Shah
> Was Abdul Abulbul Amir.
>
> —Anonymous ballad

Going through my files on Dr. Irving Joshua Matrix, the greatest numerologist in the world, I find notes on many escapades that I have not yet written about in his peripatetic career. There was the year he spent in Tübingen as founder and director of the Institute of General Eclectics, a philosophical school maintaining that all metaphysical and religious systems are in substantial agreement. (See Chapter 5 of my *Science: Good, Bad, and Bogus.*) I have never told about Dr. Matrix's revival in Bombay of phrenology, which he cleverly combined with the ancient Hindu technique of acupuncture (a method quite different from that of the Chinese). Nor have I disclosed details about his notorious Parisian brothel for dogs and cats, where the madam was a large red-haired chow from Hong Kong, and pets were given free numerological readings on Saturdays.

Perhaps someday I shall recount these sordid episodes, but now, with a heavy heart, I must speak of my visit with the wily old charlatan in April 1980. I had been in Budapest attending an international magic convention at the Duna Inter-Continental Hotel, where I had a comfortable room with a breathtaking view of the Danube. Dr. Matrix's daughter Iva somehow learned I was there. One day while I was out she telephoned, leaving the cryptic message "Jeremiah 33:3," followed by an Istanbul phone number.

The Gideon Bible in my room provided the verse: "Call unto me, and I will answer thee, and shew thee great and mighty things, which thou knowest not." Iva answered the phone when I called. Had I ever, she wanted to know, been to Istanbul? I told her I had not. She and her father would be there for a week, she said, staying at the Hilton Hotel on Taksim Square, on the European side of the ancient city.

I flew to Istanbul early the next morning, taking a room at the Santral Hotel, which is near the Hilton but considerably cheaper. When Iva came by for me at half-past ten, driving a rented American car, I was surprised to find her clad in a traditional burka of bright orange that covered everything except her hands and feet and her dark, enigmatic eyes. I was to call her Fatima, she said. Her father was in Istanbul on a top-secret mission for the U.S. Government, the nature of which she could not disclose. He had assumed the identity of a Muslim from Teheran and was using the name Abdul Abulbul Amir. Because he would not be free to meet us until late that afternoon she suggested we explore the city.

We drove south in bumper-to-bumper traffic to the accompaniment of wildly honking horns. Iva zigzagged adroitly through incomprehensible traffic lights, navigating by way of the old Jewish quarter, past the cone-capped Galata Tower and across the Galata Bridge. The murky water on each side—the Bosporus to the east and the Golden Horn inlet to the west— heaved with flotsam. The stench of sewage diminished only slightly as we moved deeper into Istanbul's Old City sector and parked near the Great Bazaar.

What bedlam! The rubbishy streets lined with tiny shops throbbed and jangled with swarms of people in every imaginable mode of dress. The more traditional of the women wore long coats and head scarfs, but some were in smart European clothes and a few even wore shorts. All of them stared at Iva in her burka as if she had been transported by a time machine

from the days of Ali Baba. As we pushed our way through the
crowds scrawny cats darted between our legs, and it seemed
that everywhere we turned there were young boys either shin-
ing shoes on luridly decorated boxes or hawking black-market
Marlboro cigarettes with shouts of "Mah-buh-ro." A strong scent
of spices almost masked the smells wafting from the surrounding waters.

Iva paused at a table of costume jewelry and after lengthy
haggling bought four inexpensive trinkets at four different
prices. One item, a pair of scarlet earrings, cost $1 in U.S.
currency. When the young shopkeeper, pretending to be angry
at the low settlement, added the four prices on his pocket
calculator, I noticed that he hit the multiplication button three
times instead of the addition button. When I pointed this out
in a whisper to Iva, she nodded but gave the man the $6.75 that
showed on the calculator's display.

"Why didn't you protest?" I asked as we elbowed our way to
another shop.

"Because," she replied, "I added the prices in my head and
they came to the same amount."

I did some scribbling on the back of an envelope. "By the
beard of the prophet!" I exclaimed. "You're right!"

Even more surprising, I later discovered that only one set of
four different prices that includes $1 has $6.75 as both its product and its sum. In the answer section I shall give the solution to
this pleasant little problem in Diophantine analysis.

We lunched at the Havuzlu restaurant near the post office,
and for the next four hours Iva took me around the city. We
visited the Blue Mosque and the Topkapi Palace. We drove
past the old Byzantine walls west of the city. It was distressing
to see how many of the beautiful mosques are decaying. Some
are used now for storing soft drinks; others house squatters.
Once-elegant mosaic walls are pockmocked with gaps where
the tiles have fallen. Even the domes and spires are stained
brown from pollution, and it was difficult to see them through

the thick daytime haze.

When we finally arrived at the Hilton, Dr. Matrix was in his suite waiting for us, wearing a striped blue suit with a small emerald crescent in his lapel. His hair was closely cropped. I assumed that his gray beard and mustache were authentic, but his piercing green eyes had been turned black by contact lenses.

"You've not been in Afghanistan, I perceive," he said as we shook hands.

"No, thank Allah," I said with a smile, recognizing Dr. Matrix's parody of Sherlock Holmes's first remark to Watson. "How on earth did you know that?"

Dr. Matrix shrugged. "My daughter keeps good track of you."

Iva excused herself to change to less cumbersome attire, and Dr. Matrix and I seated ourselves in the bedroom he was also using as an office. On his desk was a large ivory cube that had been sliced in two places and hinged so it opened to form three four-sided skew pyramids, each with a square base (as shown in Figure 35).

"The three pyramids are congruent," said Dr. Matrix. "If the square base has a side of 1, two adjacent faces are isosceles right triangles with sides of 1 and a hypotenuse that is the square root of 2. The other two sides are scalene right triangles with sides of 1 and the square root of 2, and a hypotenuse that is the square root of 3. The pyramids are easy to make with cardboard, but you would be surprised how many people have trouble putting them together to form a cube. The dissection goes back to ancient China. The pyramids were called yangmas. You might ask your readers if they can discover a completely different way to cut a cube into three identical solid forms."

Dr. Matrix picked up the hinged yangmas and folded them back until their square bases were mutually perpendicular. "Fit eight of these triplets over the eight corners of a cube of side 2," he continued, "and you create a rhombic dodecahedron. This construction provides an easy way to calculate the volume

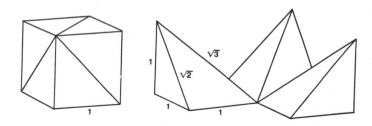

Figure 35. Dr. Matrix's cube (left) cut to form three identical skew
pyramids (right)

of such a solid. If the central cube has a side of 2, the rhombic dodecahedron has a volume of 8 + (24/3), or 16. Moreover, if you make four identical yangmas, they will fit together to form a pyramid that resembles the Great Pyramid of Egypt, with a 2-by-2 square base and four congruent isosceles triangles as sides."

The skeleton of a rhombic dodecahedron with its 12 identical diamond faces is shown at the bottom of Figure 36. The unfolded pyramid that can be made with four yangmas is shown at the top left in the illustration, and a fascinating toy can be created by gluing six of these pyramids at their bases to six square cells marked on a cross made of tape as is shown at the top right. Paint the bottom of the tape red and the sides of the pyramids blue. Folding the pyramids inward creates a solid red cube. Folding them outward creates a blue rhombic dodecahedron with a cubic interior hole. With two such models it is possible to display a blue rhombic dodecahedron, remove its "shell" to disclose an interior red cube and fold the shell to make another red cube of the same size. Each cube can then be opened into two identical blue rhombic dodecahedrons.

Each corner of Dr. Matrix's ivory cube was marked with a different digit from the set 0, 1, 2, 3, 4, 5, 6, 7. The digits were cleverly placed, he told me, so that the sum of the two

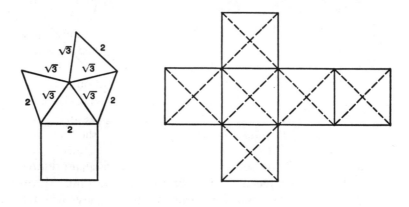

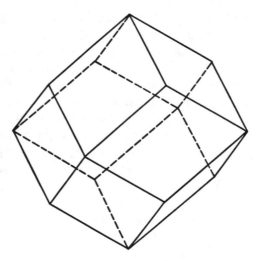

Figure 36. Plans (top) for a toy that forms both a rhombic
dodecahedron (bottom) and a cube

digits at the ends of each edge would be a prime number. (The primes are not necessarily different.) Can the reader find the only arrangement of the eight digits with this property?

"By the way," said Dr. Matrix as I jotted down the cube's number pattern, "are you aware that every cube has a volume equal to its surface area?"

He seemed mildly amused by the astonishment on my face. "Take any cube," he said, "and divide its edge into six equal parts. Call each part a hexling. A face obviously has an area of 36 square hexlings, and so since there are six faces, the total area is 6×36, or 216, square hexlings. The volume is of course 6^3, or 216, cubic hexlings. In the same way you can show that the area of any square is equal to its perimeter by dividing the side of the square into four equal quadlings. The paradox is related to a confusing proof that the surface-to-volume ratios of a sphere and a cube are the same."

"But isn't it well known that of all solids the sphere has the smallest such ratio? That's the reason soap bubbles are spherical."

"True," said Dr. Matrix, "but hear me out." He explained the "proof" as follows. If d is the diameter of a sphere, its surface is πd^2 and its volume is $(\pi d^3)/6$. The surface-to-volume ratio, then, reduces to $6/d$. Now let d be the edge of a cube. Here the surface-to-volume ratio is $6d^2/d^3$, which also reduces to $6/d$. Obviously something is wrong, but what?

"Enough of geometry," I said, my head spinning. "Have you encountered any number oddities since you came to Istanbul?"

Instead of replying, Dr. Matrix tossed over to me a 60-page booklet in English titled *Number 19: A Numerical Miracle in the Koran.* I later discovered that the author of this monograph, Rashad Khalifa, is an Egyptian who received a doctorate in biochemistry from an American university, where he also taught for a time. His booklet was published privately in the U.S. in 1972.

The number 19, Dr. Matrix pointed out, is as inscrutable to

Muslims as 666, the Number of the Beast, is to Christians. Verses 27 through 31 of Chapter 74 in the Koran tell how hell is guarded by 19 angels and explain that this number is intended to be an enigma for unbelievers. Dr. Khalifa's monograph attempts to show that 19 appears throughout the Koran too often to be there by chance. For example:

There are 114 chapters, a multiple of 19, in the Koran. A famous invocation verse called the *Basmala* ("In the name of Allah, most gracious, most merciful") is above every chapter except the ninth, but it appears a second time in the middle of Chapter 27, making 114 appearances in all. Its first word (*ism*) occurs 19 times in the Koran proper. Its second word (*Allah*) is found 2,698 (19 × 142) times. Its third word (*Al-Rahman*) appears 57 times (19 × 3), and the fourth word (*Al-Rahmin*) occurs 114 times (19 × 6).

"It's an ingenious study of the Koran," said Dr. Matrix, "but it would have been more impressive if Khalifa had consulted me before he wrote it. Nineteen is an unusual prime. For example, it's the sum of the first powers of 9 and 10 and the difference between the second powers of 9 and 10. Do you know what an emirp is?"

I shook my head.

"Well, *emirp* is *prime* backward, and it's the name my friend Jeremiah P. Farrell uses for any prime that is not a palindrome but yields a different prime when its digits are reversed. For example, the last emirpal year was 1979. The next will be 3011. Unfortunately both dates include duplicate digits. To a numerologist, emirps with no two digits alike are far more interesting. I call these numbers no-rep emirps. Their sequence starts 13, 17, 31, 37, 71, 73, 79, 97,107, ... The set is obviously finite because any number longer than ten digits must contain duplicates."

"Is there any connection between emirps and Istanbul?"

"I'm coming to that," Dr. Matrix replied. "As you know, Istanbul was once the great city of Constantinople. Its name was changed to Istanbul in 1930. Note the 19 and the 30. Nineteen is the mysterious Koran prime, and 30 is the largest integer with the property that every smaller integer relatively prime to it (no common divisor) is itself a prime. But I digress. The most important date in the history of Constantinople is of course 1453, the year the city was conquered by the Turks. Now, 1,453 not only is an emirp, it is also a no-rep emirp. Observe too that its digits add to 13, the smallest emirp."

"Have there been many no-rep emirp years since 1453?"

"There have been eleven. The last was 1879 and the next will be 3019. My good friend Leslie E. Card is the world authority on emirps. He calls them reversible primes. Card has a computer listing of all emirps under 10,000,000. There are four pairs with two digits, 11 pairs with three digits, and 42 pairs with four digits."

Card also discovered, Dr. Matrix told me, that only one six-digit emirp is cyclic, in the sense that if the first digit in the number is shifted repeatedly to the other end, each of the resulting permutations is an emirp. This unique number is 193,939. In other words, if this number is written with its digits in a circle, one can begin at any digit and go around in either direction to get a six-digit prime. There are no cyclic emirps with three, four, five, or seven digits. Is there a cyclic emirp with more than seven digits?

Card has entertained himself, Dr. Matrix said, by constructing emirp squares of digits with the property that every row, column and main diagonal is a different emirp. Thus a square of n-by-n digits would contain $4(n + 1)$ distinct primes. There is no such square of order 2 or 3. Examples for orders 4 and 5 are as follows:

```
9 1 3 3       1 3 9 3 3
1 5 8 3       1 3 4 5 7
7 5 2 9       7 6 4 0 3
3 9 1 1       7 4 8 9 7
              7 1 3 9 9
```

There are many other order-5 squares, but the order-4 square is truly extraordinary. Ignoring rotations and reflections, it is the only possible square of this type.

Can similar squares be made with no-rep emirps? No, because all primes except 2 and 5 end in 1, 3, 7, or 9: only those four digits can border an emirp square, and so if the square is of an order higher than 4, no outside prime will be free of repetitions.

I wish I had space for more of Dr. Matrix's comments about primes. He pointed out that the squares of the first seven primes add to 666, and he mentioned the even more astounding fact that if the English names for primes are alphabetized, the first number on the list is 8,018,018,851. Is the last prime on this list also determinable? Dr. Matrix thought it was, but he believed that a computer would be needed to find it.

At this point Iva, now dressed in gray silk pants and a yellow blouse, came in with a tray holding three martinis. We chatted about nonmathematical topics until the towers and domes of Istanbul became black silhouettes against a flaming gold-red sky. Far to the east, an almost invisible speck on the Black Sea's horizon, glided the ruby yacht of Omar Khayyam. It was a vision straight out of *The Arabian Nights*.

Beautiful sunsets, Iva remarked, were the only admirable by-product of Istanbul's dirty air. Through the open window, high above the city, floated the wailing of a muezzin, his call to twilight prayer amplified by loudspeakers. (Mohammed disliked bells.)

Dr. Matrix unrolled an intricately tessellated prayer rug and placed it on the floor with the point of its pattern directed southeast. After removing his shoes he recited the *Fatiha*, the first chapter of the Koran, in a loud voice. Then he knelt on the rug to prostrate himself toward Mecca while Iva sat sipping her martini with a bemused smile.

I spent several delightful days in Istanbul, and when I left, I fancied I could see tears in Dr. Matrix's eyes. Did he have a premonition about his kismet? His last words to me were *"Gule gule"* (go with laughter).

"*Salaam*," said Iva.

Three weeks later, back in New York, I was shattered by a story in the *New York Times*. It was datelined Bucharest. A Muslim known as Abdul Abulbul Amir, said to have been on a secret mission for the CIA, had met in Bucharest with a Russian agent, Ivan Skavinsky Skavar. The two had gone to a desolate spot on the delta of the blue Danube, outside the city of Izmail near the Romanian border. What happened there was unclear. Apparently the two men had fired revolvers simultaneously and both had died instantly. A peasant who had witnessed the scene from a nearby hilltop reported hearing the taller man cry *"Allah Akbar!"* as he fell.

A few words may suffice to tell the little that remains. Amir's only surviving relative, a daughter named Fatima, had arranged for her father's burial in a tomb on the bank of the Danube near the spot where he had died. It was rumored, said the *Times*, that a group of Russians had taken Skavar's body aboard a ship and disposed of it in the Black Sea. No doubt I will learn more details if and when I see Iva again. With these sad words I close my final account of him whom I shall ever regard as the strangest and the wisest man I have ever known.

Answers and Commentary

One

I. The letters *OTTFFSSENT* are the initials of the names of the cardinal numbers from 1 to 10. Mrs. Georgianna March, an editor of the *Bulletin of the Atomic Scientists*, pointed out in a letter that if the first and last letters are transposed, making the order *TTTFFSSENO*, they are the initial letters of the first ten multiples of 10, from 10 to 100 (one hundred).

II. Dr. Matrix's addition problem was originated by Alan Wayne, a high-school teacher of mathematics in New York, and first appeared in the *American Mathematical Monthly* (Aug.-Sept. 1947, p. 413). In introducing the problem, the magazine's problem editor pointed out that a "cryptarithm," to be considered "charming," should exhibit four features:

The letters should make sense.
All digits should be used.
The solution must be unique.
It should be solvable by logic rather than by tedious trial and error.

Wayne's cryptarithm has all four features. The unique solution is:

$$29786$$
$$850$$
$$850$$
$$\overline{31486}$$

Note that the sum differs in only one digit from the four-decimal value of pi.

For readers who may wonder how to go about solving a cryptarithm, I quote a letter of Monte Dernham, of San Francisco, who sent me the best explanation of how Wayne's problem could be analyzed:

The repetition of TY in the first and fourth lines necessitates zero for N and 5 for E, with unity carried to the hundreds column. The double space preceding each TEN requires that O in FORTY equal 9, with 2 carried from the hundreds column, whence I denotes the unit digit 1 in 11, with F plus 1 equal to S. This leaves 2, 3, 4, 6, 7, and 8 unassigned.

Since the hundreds column (viz., R plus 2T plus 1) must be equal to or greater than 22, T and R must each be greater than 5, relegating F and S to 2, 3, and 4. Now X is not equal to 3; else F and S could not be consecutive integers. Then X equals 2 or 4, which, it is readily found, is impossible if T is equal to or less than 7. Hence T equals 8, with R equal to 7 and X equal to 4. Then F equals 2 and S equals 3, leaving the remaining letter, Y, equal to 6.

For another excellent cryptarithm by Wayne, see chapter 18.

Two

I. The only other rookwise-connected antimagic square is the "complement" of the one in Figure 2. Simply change each digit to the difference between that digit and 10. The result is a square that can be obtained by spiraling the digits in the same way as before, but taking them in reverse order:

987
216
345

One way to prove there are no other such squares starts by observing that if the matrix is colored like a chessboard, with white cells at the corners, the odd digits must be on white, even digits on black. Digits 2 and 4 cannot be opposite because 3 would have to go between, and this makes a rook path impossible. Either 8 or 6, therefore, must be opposite 2. The path must start and end on white. It takes only a few minutes to check the four essentially different patterns for duplication of sums.

I had not seen an antimagic square before Dr. Matrix introduced me to them. The earliest example I later found of such a square is the order-3 square given in *Sam Loyd and His Puzzles* (1928) as the answer to a puzzle on page 44.

In *Mathematics Magazine* (Jan. 1951) Dewey Duncan defined a heterosquare as a square in which no two rows, columns, or diagonals (including "broken diagonals") have the same sum. (The order-3 square has four broken diagonals. Referring to the square shown in Figure 2, they are the cells bearing the number triplets 1, 6, 4; 8, 2, 5; 3, 8, 6; and 2, 4, 7. Thus the antimagic square in Figure 2 is not a heterosquare; for one thing, the third broken diagonal adds up to 17, and so does the second column.) Duncan asked for a heterosquare of order-3 and proof that no such square of order-2 exists. It is easy to show that an order-2 is impossible. A proof that the order-3 also is impossible was given by Charles F. Pinzka in *Mathematics Magazine* (Sept.-Oct. 1965, pp. 250–252). Order-4 squares *are* possible; Pinzka gave two. Another proof of impossibility for the order-3 was given by Prasert Na Nagara in the same magazine (Sept.-Oct. 1966, pp. 255–256). Nagara also found two "almost" heterosquares of order-3 in which all sums but two were distinct.

J. A. Lindon, writing in *Recreational Mathematics Magazine* (Feb. 1962), proposed searching for antimagic squares in which the sums of the rows, columns, and main diagonals (broken diagonals not considered) are not only different but form a sequence of consecutive integers. A summary of Lindon's results, with some new material added, appears in Joseph Madachy's *Mathematics on Vacation* (New York: Scribner, 1966), pp. 101–110. No order-2 square of this type is possible. Order-3 also is impossible, although one can come close, as the following square (from C. C. Verbeek's *Puzzel met Plezier*, Amsterdam, 1962, p. 155) shows:

268
791
534

All eight sums are distinct, and only one diagonal sum, 22, is outside the sequence.

Many order-4 and higher antimagic squares, with all sums in consecutive order, were found by Lindon.

Charles W. Trigg, writing on "The Sums of Third Order Anti-Magic Squares," *Journal of Recreational Mathematics* 2 (1969):250–254, showed that the eight sums of an order-3 antimagic square cannot be in any arithmetic progression, thus confirming Lindon's conjecture that they cannot be consecutive. He also proved that the eight sums cannot all be even.

In a note on "A Remarkable Group of Antimagic Squares," *Mathematics Magazine* 44 (1971):13, Trigg examined the eight patterns obtained by placing 1 in the center of the three-by-three array, the sequence 3, 5, 7, 9 in the corners, and the sequence 2, 4, 6, 8 in the side cells. "Remarkably, whether the sequences run clockwise or counterclockwise, each of the eight essentially distinct squares thus obtained are antimagic."

The complements of these eight squares are also antimagic. When the four broken diagonals are considered, it turns out that each of the sixteen squares is also "almost heterosquare" in having only two duplicate sums. Commenting on problem 84 in the same magazine, 4 (1971):236–237. Trigg has given a method that produces 108 order-3 arrays that are almost heterosquare. The total number of distinct order-3 antimagic squares, and the number of distinct order-3 almost heterosquares, remain unknown.

II. The equation asked for is:

$$36^2 + 37^2 + 38^2 + 39^2 + 40^2 = 41^2 + 42^2 + 43^2 + 44^2$$

I am indebted to Russell L. Linton, Oakland, California, for pointing out in a letter that the first integer in the series of such equations is obtained by the formula $n(2n+1)$, where n is the number of terms on the right side of the equation. Thus, to write the next example, which has five terms on the right, we substitute 5 for n to obtain $5(10+1)=55$. We can immediately write:

$$55^2 + 56^2 + 57^2 + 58^2 + 59^2 + 60^2 = 61^2 + 62^2 + 63^2 + 64^2 + 65^2$$

A discussion of this series, "Runs of Squares," by T. H. Beldon, appeared in the *Mathematical Gazette* (Dec. 1961, pp. 334–335).

The series has a trivial analogy with the following first-power series:

$$1 + 2 = 3$$
$$4 + 5 + 6 = 7 + 8$$
$$9 + 10 + 11 + 12 = 13 + 14 + 15$$

III. The three-symbol chain problem has a fascinating history that begins with a two-symbol chain first discovered by the Norwegian mathematician Axel Thue and described by him in 1912. Begin with 01. For the 0, substitute 01, and for the 1, substitute 10. The result is a chain of four digits: 0110. Repeating this procedure, changing each 0 to 01 and each 1 to 10, produces the chain 01101001. In this way we can form a chain as long as we wish, each step doubling the number of digits and forming a chain that starts by repeating the previous chain. This sequence of symbols, called the Thue series, has the remarkable property that no block of one or more digits ever appears three times consecutively. The chain may "stutter" once, but whenever this occurs, regardless of the size of the block that repeats, the very next digit is sure to be the wrong one for a third appearance of the block.

Max Euwe, a former world chess champion, was among

the first to recognize that the Thue sequence provides a method of playing an infinitely long game of chess. The so-called German rule for preventing such games declares a game drawn if a player plays any finite sequence of moves three times in succession in the same position. Two players need only create a position in which each can move either of two pieces back and forth, regardless of how the other player moves his two pieces. If each now plays his two pieces in a Thue sequence, neither will ever repeat a pattern of moves three times consecutively.

From the Thue series it is easy to derive a three-symbol chain that solves Dr. Matrix's problem. First, we transform it to a chain of four symbols by writing 0 under every 00 pair, 1 under every 01 pair, 2 under every 10 pair, and 3 under every 11 pair:

Thue series:	0 1 1 0 1 0 0 1 . . .
Four-symbol chain:	1 3 2 1 2 0 1 . . .

This infinite four-symbol chain has the property that no finite block of digits ever appears *twice* side by side. It can now be transformed to a three-symbol chain, with the same property, by replacing every 3 with a 0:

Four-symbol chain:	1 3 2 1 2 0 1 . . .
Three-symbol chain:	1 0 2 1 2 0 1 . . .

This solution to the three-symbol problem was given by Marston Morse and Gustav Hedlund in an important 1944 paper, "Unending Chess, Symbolic Dynamics and a Problem in Semigroups," *Duke Mathematics Journal* 11 (1944):1–7. There were earlier solutions (including one by the Russian mathematician S. Arshon in 1937) and many later ones. John Leech gave this solution in "A Problem on Strings of Beads," *Mathematical Gazette* 41 (1957):277–278:

Consider the following three blocks of digits:

0 1 2 1 0 2 1 2 0 1 2 1 0

1 2 0 2 1 0 2 0 1 2 0 2 1

2 0 1 0 2 1 0 1 2 0 1 0 2

The digits in these blocks are so arranged that if we substitute the three blocks for the three digits (replacing 1 with one block, 2 with another, 3 with the third) in any stutter-free chain (e.g., any one of the three blocks), the resulting chain will also be stutter-free. In this longer chain we can now substitute blocks for digits once more to obtain a still longer chain, and so on ad infinitum.

It is not possible to construct shorter palindromic blocks (blocks that are the same backward as forward) that can be used in this way, but shorter asymmetric blocks are possible. Allan Beek sent me a similar solution using the following asymmetric blocks of eleven digits each:

12313231213
12321312132
12321323132

It is not known if there is a set of three shorter blocks that provides a proof of this type.

The three-symbol chain furnishes a way of evading the rule for drawn chess games even if the rule is strengthened by declaring a game drawn if a finite sequence of moves occurs only twice in succession. Each player simply moves three pieces in a pattern given by the three-symbol chain.

There are other ways of generating the Thue series than the one explained above. In 1961 Dana Scott sent me the following. First write the sequence of integers in binary

form: 0, 1, 10, 11, 100, 101, 110, 111, 1000 . . . Next re-
place each number with 1 if it contains an odd number of
1's, and with 0 if it contains an even number of 1's. The
result, surprisingly, is the Thue series: 011010011 . . .

A method of transforming the Thue series directly to a
three-symbol solution of Dr. Matrix's problem was explained in 1963 by C. H. Braunholtz, "An Infinite
Sequence of Three Symbols with No Adjacent Repeats,"
American Mathematical Monthly 70 (1963):675–676. In
the Thue series the number of 1's between any 0 and the
next 0 is either 0, 1, or 2. There are two 1's between the
first and second 0, one 1 between the second and third 0's,
none between the third and fourth, and so on. The
numbers of these 1's, as we proceed from 0 to 0, form a
three-symbol infinite series, 2102012 . . . , with the
required property.

P. Erdös proposed the following three-symbol chain
problem that is the same as the one given by Dr. Matrix
except that two blocks of digits are now considered "identical" if each symbol appears in them the same number of
times. For example, 00122 = 02102 because each contains
two 0's, one 1, and two 2's. The largest possible sequence
that does not have two "identical" blocks side by side is
one of seven digits, e.g., 0102010. It is not yet known if
there is an infinite four-symbol chain with this property.

Other references on the Thue series and the three-symbol problem include:

Marston Morse, "A Solution of the Problem of Infinite
Play in Chess," Abstract 360, *Bulletin of the American
Mathematical Society* 44 (1938):632.

D. Hawkins and W. E. Mientka, "On Sequences Which
Contain No Repetitions," *Mathematics Student* 24
(1956):185–187.

G. A. Hedlund and W. H. Gottschalk, "A Characterization of the Morse Minimal Set," *Proceedings of the American Mathematical Society* 15 (1964):70–74.

Richard A. Dean, "A Sequence Without Repeats," *American Mathematical Monthly* 72 (1965):383–385.

P. A. B. Pleasants, "Non-Repetitive Sequences," *Proceedings of the Cambridge Philosophical Society* 68 (1970):267–274.

T. C. Brown, "Is There a Sequence of Four Symbols in Which No Two Adjacent Segments Are Permutations of One Another?" *American Mathematical Monthly* 78 (1971):886–888.

R. C. Entringer, D. E. Jackson, and J. A. Schatz, "On Nonrepetitive Sequences," *Journal of Combinatorial Theory*, Series A, 16 (1974):159–164.

IV. The number 102564 quadruples in size if the 4 is moved from the back to the front, 410256; therefore, Miss Toshiyori's telephone number is 1–0256. Puzzles of this type are easily solved by a kind of multiply-as-you-go technique explained in Figure 37.

After mastering this method, readers may wish to tackle the following three problems:

1. What is the smallest number ending in 6 that becomes six times as large when the 6 is shifted from the end to the front? (Warning: The number has 58 digits!)

2. Find the smallest number *beginning* with 2 that triples when the 2 is moved to the end.

3. Prove that there is no number beginning with the digit n that increases n times when the first digit is moved from the front to the end, except in the trivial case where n is 1.

Readers interested in further explorations of problems of this type, in which digits are moved from one end of a

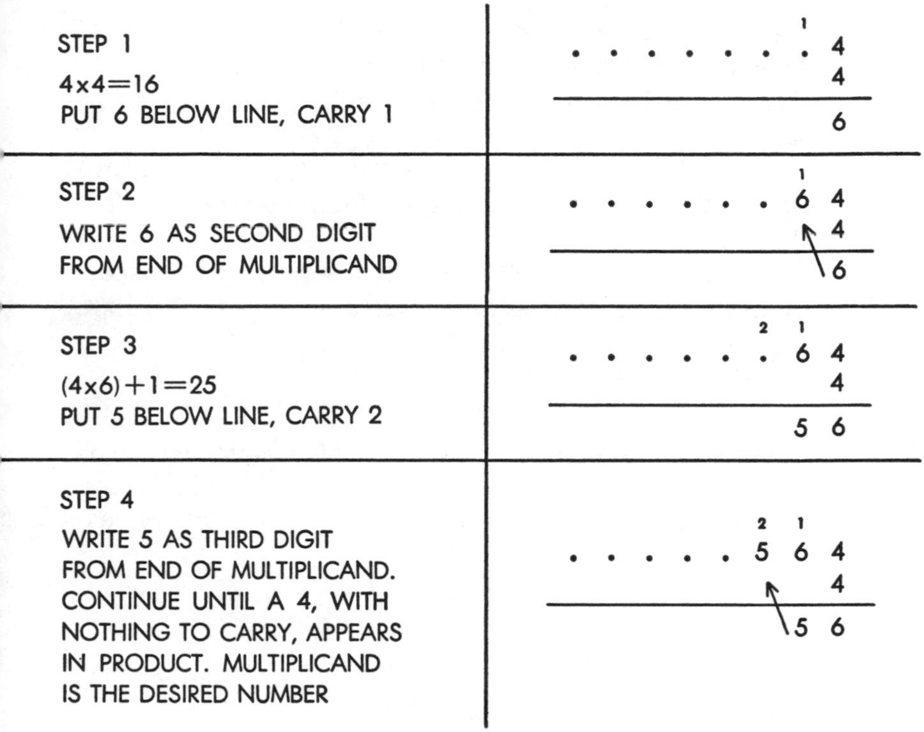

STEP 1 4×4=16 PUT 6 BELOW LINE, CARRY 1	$\overset{1}{\ldots\ldots\ldots} 4$ 4 $\overline{6}$
STEP 2 WRITE 6 AS SECOND DIGIT FROM END OF MULTIPLICAND	$\overset{1}{\ldots\ldots 6}\, 4$ 4 $\overline{6}$
STEP 3 (4×6)+1=25 PUT 5 BELOW LINE, CARRY 2	$\overset{2\ \ 1}{\ldots\ldots 6}\, 4$ 4 $\overline{5\ \ 6}$
STEP 4 WRITE 5 AS THIRD DIGIT FROM END OF MULTIPLICAND. CONTINUE UNTIL A 4, WITH NOTHING TO CARRY, APPEARS IN PRODUCT. MULTIPLICAND IS THE DESIRED NUMBER	$\overset{2\ \ 1}{\ldots\ldots 5}\, 6\, 4$ 4 $\overline{5\ \ 6}$

Figure 37. How to solve the problem of Miss Toshiyori's telephone
       number

number to another to accomplish specified results, will
find helpful the following references:

Aaron Bakst, *Mathematical Puzzles and Pastimes* (New
York: Van Nostrand, 1954), pp. 177f.

L. A. Graham, *Ingenious Mathematical Problems and
Methods* (New York: Dover, 1959), problem 72.

Dan Pedoe, *The Gentle Art of Mathematics* (New York:
Macmillan, 1958), pp. 11f.

W. B. Chadwick, "On Placing the Last Digit First," *American Mathematical Monthly* 48 (1941):251.

D. E. Littlewood, "A Digit Problem," *Mathematical Gazette* 39 (1955):58.

W. D. Skees, "A Permutative Property of Certain Multiples of the Natural Numbers," *Fibonacci Quarterly* 3 (1965):279f.

Charles W. Trigg, "Division of Integers by Transposition," *Journal of Recreational Mathematics* 1 (1968):180f.

Joseph S. Madachy, "A Fibonacci Constant," *Fibonacci Quarterly* 6 (1968):385f.

When Sin Hitotumato translated this chapter into Japanese, he pointed out that Iva's phone number must satisfy the equation $4 \times 10^5 + x = 4 (10x + 4)$. This gives x a value of $399984/39 = 10256$.

Three

I. A fraction has the repeating decimal form .272727 . . . What is the fraction?

Write the equation:

$$x = .272727 \ldots$$

Multiply both sides by 100:

$$100x = 27.272727 \ldots$$

Subtract the first equation:

$$100x = 27.272727 \ldots$$
$$\underline{x = .272727 \ldots}$$
$$99x = 27$$

We conclude:

$$x = {}^{27}/_{99} = {}^{3}/_{11}$$

In general, the procedure is to write the repeating block of decimals above the line, put the same number of 9's below, then reduce the fraction to lowest terms.

II. Only four numbers equal the sum of the cubes of their digits: 153, 370, 371, and 407.

G. H. Hardy mentions those four numbers in his famous little book *A Mathematician's Apology,* and adds: "These are odd facts, very suitable for puzzle columns and likely to amuse amateurs, but there is nothing in them which appeals to the mathematician."

Hardy was right, of course. But most numerologists *are* amateur mathematicians, and the general problem—finding numbers of *n* digits that equal the sum of the *n*th powers of their digits—continues to have its fascination. Many readers with access to computers searched for higher numbers of this type. Here is a summary of the most up-to-date results known to me:

Order 1. Every positive integer is, of course, equal to itself.

Order 2. No integer is equal to the sum of the squares of its digits, except in the trivial case of 1. A proof by N. J. Fine is given in *American Mathematical Monthly* (Nov. 1964, pp. 1042–1043).

Order 3. Four numbers (given above) equal the sum of the cubes of their digits. All are three-digit numbers.

Order 4. Three numbers, all with four digits, equal the sum of the fourth powers of their digits:

$$1,634$$
$$8,208$$
$$9,474$$

Order 5. Three five-digit numbers equal the sum of the fifth powers of their digits:

$$54,748$$
$$92,727$$
$$93,084$$

There are also two four-digit numbers:

4,150
4,151

And one six-digit number:

194,979

Order 6. Surprisingly, only one number—and it has six digits—equals the sum of the sixth powers of its digits:

548,834

Order 7. Four seven-digit numbers equal the sum of the seventh powers of their digits:

1,741,725
4,210,818
9,800,817
9,926,315

And one eight-digit number:

14,459,929

Order 8. Three numbers, all with eight digits, equal the sum of the eighth powers of their digits:

24,678,050
24,678,051
88,593,477

Order 9. Four numbers, all with nine digits, equal the sum of the ninth powers of their digits:

146,511,208
472,335,975
534,494,836
912,985,153

Order 10. One number, believed to be unique, equals the sum of the tenth powers of its digits:

$$4,679,307,774$$

This number was discovered by Harry L. Nelson, who published it in his report "More on PDI's" (Publication UCRL-7614, University of California, Dec. 1, 1963). Nelson's results are summarized in Joseph Madachy's *Mathematics on Vacation* (New York: Scribner, 1966), p. 164, in a section dealing with this and related problems.

An integer of n digits which is equal to the sum of the kth powers of its digits is now called (following the terminology of Max Rumney) a PDI (perfect digital invariant); if $n = k$, it is called a PPDI (pluperfect digital invariant). Both belong to an ill-defined genus of "self-generating" or "narcissistic" numbers.

It is not yet known if the number of PDI's is finite or infinite, or if a PDI can be prime.

The number of PPDI's is finite. This seems to have been first proved by D. St. P. Bernard. He observed that the maximum sum of the nth powers of n digits is $n(9^n)$. Let $n = 61$. Since $61(9^{61})$ is less than 10^{60}, it follows that no PPDI can have more than sixty digits. See Max Rumney, "Digital Invariants," *Recreational Mathematics Magazine* (Dec. 1962), pp. 6–8.

A complete statement of Bernard's proof is given by Benjamin L. Schwartz in "Finiteness of a Set of Self-Generating Integers," *Journal of Recreational Mathematics* 2 (1969):79–83. Schwartz sharpens this result by showing that no PPDI can have more than fifty-nine digits. See also the same author's "Finite Bounds on Digital Invariants—Some Conjectures," *Journal of Recreational Mathematics* 3 (1970):88–92; and "Self-Generating Integers," *Mathematics Magazine* 46 (1973):158–160.

The upper bound of 59 has been lowered by H. L. Nelson to 58, but the largest PPDI is far from known. Recent references include Joseph S. Madachy, "Some New Narcissistic Numbers," *Fibonacci Quarterly* 10 (1972):295–298; Victor G. Feser, "Narcissistic Numbers," *Pi Mu Epsilon Journal* 5(1973):409–414.

III. Dr. Matrix's cell number is 45. When a decimal point is placed between the digits, it becomes 4.5, the average of 4 and 5. The answer is unique.

IV. The floor of Dr. Matrix's cell is three by six yards. The only other rectangle of integral sides, with a perimeter equal to area, is the four by four, but Dr. Matrix specifically stated that the floor was not square.

The problem has historical interest. B. L. van der Waerden, in his beautiful book *Science Awakening* (Oxford, 1961), quotes the following passage from Plutarch: "The Pythagoreans also have a horror for the number 17. For 17 lies exactly halfway between 16, which is a square, and the number 18, which is the double of a square, these two being the only two numbers representing areas for which the perimeter [of the rectangle] equals the area."

The problem yields readily to simple Diophantine analysis. Let x and y be the rectangle's sides. The area, xy, equals the perimeter, $2x + 2y$. When written like this:

$$y = 2 + \frac{4}{x - 2}$$

it is apparent that y is integral only if x is 3, 4, or 6. This leads to the two possible answers.

V. The letters of CHESTY can be rearranged to make only one other word: SCYTHE.

VI. Iva Toshiyori's remark ("The day before yesterday I was twenty-two, but next year I'll be twenty-five") makes sense only if she made it on January 1 and her birthday is December 31.

Five

I. To change the multiplication square to a division square merely exchange each corner number with the one diagonally opposite. The result is

$$
\begin{array}{ccc}
3 & 1 & 2 \\
9 & 6 & 4 \\
18 & 36 & 12
\end{array}
$$

This is the only way (except for rotations and reflections) that the nine numbers can be arranged to form a division square. The magic division constant is 6. Multiply the end numbers of any line of three, divide by the middle number, and the result is 6. No smaller constant for an order-3 division square is possible.

Curiously, when this same procedure is applied to the familiar order-3 magic addition square, whose rows, columns, and main diagonals add to 15:

$$
\begin{array}{ccc}
8 & 1 & 6 \\
3 & 5 & 7 \\
4 & 9 & 2
\end{array}
$$

the matrix is transformed to a subtraction square:

$$
\begin{array}{ccc}
2 & 1 & 4 \\
3 & 5 & 7 \\
6 & 9 & 8
\end{array}
$$

When the end digits of any line of three are added and the middle digit subtracted, the result is 5.

For more on multiplying magic squares, sometimes called geometric magic squares, see:

Harry A. Sayles, "Geometric Magic Squares and Cubes," *Monist* 23 (1913):631–640.

Henry E. Dudeney, *Amusements in Mathematics* (London: Thomas Nelson, 1917), pp. 124f. (There is a Dover paperback reprint.)

Walter W. Horner, "Addition-Multiplication Magic Squares," *Scripta Mathematica* (Sept. 1952), pp. 300–303.

Boris Kordemsky, "Geometric Magic Squares," *Recreational Mathematics Magazine* (Feb. 1963), pp. 3–6.

Jack Gilbert, "Minimum Multiplying Magic Squares," *Mathematics Teacher* (May 1960), pp. 325–331.

II. The number 19 can be expressed with four 4's, aided by arithmetical signs and the decimal point only, as follows:

$$
\frac{4 + 4 - .4}{.4} = 19
$$

Note that this formula also gives 19 when *any number whatever* is substituted for 4. This is true of many of the other four-4's formulas given in chapter 5 and suggests an unusual generalization of the pastime: How many positive integers, starting with 1, can be expressed with four n's under the traditional limitations of permissible symbols? (See references in III below.)

III. The simplest methods I know for expressing 64 with four 4's, three 4's, and two 4's are shown below:

$$(4+4)\,(4+4)=64$$

$$4\times4\times4=64$$

$$\sqrt[\sqrt{4}]{4^{4!}}=64$$

Ruth Ann Schiller, copy editor of the *Journal of Mathematical Physics*, was the first to send this attractive method of expressing 64 with two 4's:

$$4!!\times4!!=64$$

The double factorial sign "!!" is an esoteric but standard symbol. When $2n$ is followed by "!!" it expresses the product of $2\times4\times6\times\ldots\times2n$. Thus $4!!=2\times4=8$.

A reader who did not identify himself did it this way:

$$\not{\!4}$$

Other methods of expressing 64 with two 4's can be found in:

J. A. Tierney, problem E631, *American Mathematical Monthly* 52(1945):219.

J. A. Tierney, "64 Expressed by Two Fours," *Scripta Mathematica* 18 (1952):218.

C. W. Trigg, "The Number 64 Expressed by Two Fours," *Scripta Mathematica* 19 (1953):242.

On 64 with three 4's, see:

H. S. M. Coxeter, "Rouse Ball's Unpublished Notes on Three Fours," *Scripta Mathematica* 18 (1952):85–86.

On 64 with four 4's, see:

Marjorie Bicknell and Verner E. Hoggatt, "64 Ways to Write 64 Using Four 4's," *Recreational Mathematics Magazine* (Jan.-Feb. 1964), pp. 13–15.

Are there ways to write 64 with *one* 4? Yes, by adopting suitable notation, any positive integer probably can be expressed with one 4. A method using only brackets, square-root signs, and factorials was explained by Donald E.

Knuth in "Representing Numbers Using Only One 4," *Mathematics Magazine* 37 (1964):308–310. See also S. Guttman, "The Single Digit 4," *Scripta Mathematica* (March 1956), p. 78.

Readers interested in ways of expressing positive integers above 20 with four 4's will find a list of expressions for numbers 1 to 100 in L. Harwood Clarke, *Fun with Figures* (London: Heinemann, 1954), pp. 51–53; and Angela Dunn, *Mathematical Bafflers* (New York: McGraw-Hill, 1964), pp. 3–8. A list from 1 to 30 appears in S. I. Jones, *Mathematical Wrinkles* (Jones, 1912), p. 217. A way of expressing pi with four 4's is given by J. H. Conway and M. J. T. Guy, "Pi in Four 4's," *Eureka* (Oct. 1962), pp. 18–19.

Obviously one can also waste vast amounts of time trying to express integers with two 2's, three 3's, five 5's, and so on. Many articles along these lines have been published, but I find the topic so boring that I cannot bring myself even to type out the references I have on file.

The more general task of expressing integers with four n's (formulas in which any integer can be substituted for n) is discussed by Seymour Krutman in *Scripta Mathematica* 13 (1947):47; and Donald C. B. Marsh, ibid. 15 (1949):91. Verner Hoggatt and Leo Moser, "A Curious Representation of Integers," *American Mathematical Monthly* 57 (1950):35, show how to express any integer by using any integer greater than 1, repeating it any number of times greater than three, and employing a finite number of operator symbols.

Six

I. The only four-digit number that is self-replicating is 7,641. When a number is reversed and the smaller version is subtracted from the larger, the resulting digits must sum to a multiple of 9. If the reader knows this, his search for the four-digit number is considerably narrowed, since only four digits summing to 9, 18, or 27 need be considered.

For a related diversion, put down any four digits, provided not all are alike, to form a four-digit number from 0,001 through 9,998. Arrange the four digits in descending order. Reverse the order, making it ascending, and obtain the difference between the two numbers. Repeat the procedure with the difference. You will find that after eight or fewer repetitions you will reach 6,174, which generates itself.

Initial 0's must be preserved throughout. For example: $1,000 - 0,001 = 0,999$; $9,990 - 0,999 = 8,991$; $9,981 - 1,899 = 8,082$; $8,820 - 0,288 = 8,532$; $8,532 - 2,358 = 6,174$.

A similar procedure with three digits, not all alike, quickly converges on the unique self-replicating 495. Numbers with five or more digits terminate in more than one loop. By "loop" is meant a sequence of one or more numbers which repeat cyclically.

The number 6,174 is called Kaprekar's constant after Dattatraya Ramchandra Kaprekar, of Devlali, India, who first announced its significance in "Another Solitaire Game," *Scripta Mathematica* 15 (1949):244–245. See also Kaprekar's "An Interesting Property of the Number 6174," ibid. 21 (1955):304, and his booklet *The New Constant 6174*, published by himself in 1959. His analysis of the five-digit case is given in his 1963 booklet *The New Recurring Circulating Constants from All the Five Digital Integers*.

Later articles on the Kaprekar routine, some generalizing it to other bases, include:

J. H. Jordan, "Self Producing Sequences of Digits," *American Mathematical Monthly* 71 (1964):61–64.

"Kaprekar's Constant," solution to problem E2222, *American Mathematical Monthly* 78 (1971):197–198.

Charles W. Trigg, "Predictive Indices for Kaprekar's Routine," *Journal of Recreational Mathematics* 3 (1970): 245–254; "Kaprekar's Routine with Two-Digit Integers," *Fibonacci Quarterly* 9 (1971):189–193; "Kaprekar's Routine with Five-Digit Integers," *Mathematics Magazine* 45 (1972):121–129; and "All Three-Digit Integers Lead to . . . ," *Mathematics Teacher* 67 (1974):41–45.

W. A. Tillick, of Lower Hutt, New Zealand, wrote me that he had programmed a computer to search for all numbers of four distinct digits, *not* in descending sequence, that are self-replicating. There are five:

$$9,108 - 8,019 = 1,089$$
$$5,823 - 3,285 = 2,538$$
$$3,870 - 0,783 = 3,087$$
$$2,961 - 1,692 = 1,269$$
$$1,980 - 0,891 = 1,089$$

Two readers, Lieutenant Bernard F. Shearon, Jr., and Stephen E. Payan, wrote to point out that the self-replicating number 987,654,321 has an analog in all number systems with even bases. In an octal system, for example, $7,654,321 - 1,234,567 = 6,417,532$. In a 12-based system (letting $x = 10$, $y = 11$):

$$yx,987,654,321$$
$$12,345,678,9xy$$
$$\overline{x8,641,y97,532}$$

II. The hotel room number is 497. This number plus 2 is 499; multiplied by 2, the result is 994. The integers 497 and 2 are the only pair that, when multiplied and added, yield two three-digit numbers, one the reverse of the other.

There are three pairs of numbers that behave in the same way to give two-digit results: $24 + 3 = 27$, $24 \times 3 = 72$; $47 + 2 = 49$, $47 \times 2 = 94$; $9 + 9 = 18$, $9 \times 9 = 81$.

III. These are the only possible ways to apply five or fewer plus or minus signs to the ascending and descending series so that the sum is 65:

$$123 + 4 - 56 - 7 - 8 + 9 = 65$$
$$-98 + 76 + 54 + 32 + 1 = 65$$

IV. One way to place four marks on a twelve-inch ruler so that all integral lengths from 1 through 12 can be measured directly is to place the marks at 1, 4, 7, and 10 inches from one end.

It is easy to prove that ten is the maximum number of different lengths that can be measured by a ruler with only three marks. If the three marks and the ruler's two ends are numbered by their distance from 0 at one end, every measurable length corresponds to a difference between two of the five numbers. The number of different ways to pair five distinct numbers is $4 + 3 + 2 + 1 = 10$, so clearly ten is the maximum. One way to place three marks on a twelve-inch ruler so that ten different lengths can be measured is to place the marks at 1, 4, and 10. It is not possible to place three marks on a ten-inch ruler so that all ten lengths can be measured.

Eight marks, the minimum, can be placed on a yardstick in such a way that all integral lengths from 1 through 36 can be measured. They are placed at 1, 3, 6, 13, 20, 27, 31, and 35. This solution was first found by John Leech, who gave it in his classic paper on this problem, "On the Representation of $1, 2, \ldots, n$ by Differences," *Journal of the London Mathematical Society* 31 (1956):160–169.

Henry E. Dudeney, in problem 180, *Modern Puzzles* (1926), asked for the placing of eight marks on a thirty-three-inch ruler and gave sixteen solutions. Dudeney undoubtedly believed that at least nine marks were required for any ruler longer than thirty-three inches. This was believed to be the case until Leech extended it to thirty-six inches, the maximum for eight marks (or nine segments). It also has been established, by J. C. P. Miller, that Leech's solution is the only one. The general problem is still far from solved, leading into many thorny related problems and deep aspects of combinatorial arithmetic.

Leech's results in his 1956 paper have been extended by C. B. Haselgrove and Leech in their "Note on Restricted Difference Bases," *Journal of the London Mathematical Society* 32 (1957):228–231; and by B. Wickmann's note, ibid. 38 (1963):465–466. See also T. H. O'Beirne's comments on the ruler problem in *Puzzles and Paradoxes* (London: Oxford, 1965), pp. 110–111.

(I would like to thank John Bales for the Dewey decimal item, Donald C. Rehkopf for the pi pattern, Martin Cohen for the division of 987,654,321 by its reversal, and Don T. Hastings for problems involving the insertion of signs in the nine-digit series.)

Seven

I. What could be a simpler way to half-bake a pi than to add the ten digits, in the order shown, to pi's first ten decimals?

$$3.1415926535$$
$$.1234567890$$
$$\overline{3.2650494425}$$

II. The two numbers formed by the ten digits that have the largest product are 96,420 and 87,531. Among pairs of numbers that have the same sum the pair with the largest product is the pair most nearly equal. This rule leads us to solve the problem by taking the digits in descending order, starting two numbers with 9 and 8, then annexing the remaining digits by pairs, the larger of each pair always joining the smaller of the two numbers: 9 and 8, 96 and 87, 964 and 875, and so on.

III. The number 8,549,176,320, produced by the indexer named Betty, is simply the ten digits in alphabetical order. (Note that the letters in BETTY are also alphabetized.)

IV. Many mathematical magic tricks as well as off-color number stunts have been based on the principle involved in Dr. Matrix's numerological advice to Smith. The following simple proof that the described operations must result in 1,089 is given by Wallace Lee in *Math Miracles* (Durham, N.C.: privately printed, 1950), pp. 26–27:

Let *HTU* represent the digits at the hundreds, tens, and units positions in the larger of the first two numbers. Its reversal, *UTH*, cannot be subtracted without borrowing 1 from the upper *T* (which means adding 10 to the upper *U*) and since the upper *T* is now smaller than the lower *T*, a 1 must be borrowed from the upper *H* (which means adding 10 to the upper *T*). The situation looks like this:

$$\begin{array}{ccc} H-1 & T-1+10 & U+10 \\ U & T & H \\ \hline \end{array}$$

The subtraction results in:

$$\begin{array}{ccc} H-1-U & 9 & U+10-H \end{array}$$

This number is then reversed and added:

$$\begin{array}{ccc} H-1-U & 9 & U+10-H \\ U+10-H & 9 & H-1-U \\ \hline 10 & 8 & 9 \end{array}$$

My application of this to Mr. Smith and his health problem first appeared in *Ibidem*, a Canadian magic journal, no. 7 (Sept. 1956).

V. Harry Hażard's cryptarithm multiplication problem:

$$\begin{array}{r} LYNDON \\ \times \quad B \\ \hline JOHNSON \end{array}$$

has the unique answer:

$$\frac{\begin{matrix} 570140 \\ 6 \end{matrix}}{3420840}$$

Two readers, D. Gospodnetil and Douglas G. Russell, each gave this problem to a computer, which proved in a few minutes (not counting the hours of programming) that the solution was unique. Russell also programmed the computer to test for a unique solution of a problem that independently occurred to another reader, Edward C. Devereux:

$$\frac{\begin{matrix} .MARTIN \\ .A \end{matrix}}{.GARDNER}$$

The dots are decimal points. Not only does this have a unique solution, but no other middle initial can be substituted for A. to give a pattern with a solution. It would be pleasant to report that my middle initial is A.; unfortunately I have no middle name.

My report on Dr. Matrix's statement that 4 was the only honest number in English provoked a lot of mail. The doctor had in mind, of course, number names uncontaminated by operational phrases. Without this restriction an infinity of honest numbers can be found. "How about two cubed?" asked Morris H. Woskow, or "twelve plus one," "twenty minus five," "minus four squared," and so on. Walter Erbach sent a list of forty examples that included "one half of thirty," "square root of nine hundred sixty-one," and "integral of x dx from sixteen to eighteen." Michael Burke and Norman Buchignani suggested such exotic phrases as "the largest prime less than thirty" and "the odd integer between thirty-eight and forty." Paul C. Hoell came up with a 62-letter phrase, "cube root of two hundred thirty-eight thousand three hundred twenty-eight," but this was topped by Robert B. Pitkin's 101-letter phrase, "Seventh root of one zero seven trillion two thirteen billion five thirty-five million two ten thousand seven hundred one."

Two correspondents, W. M. Woods and Malcolm R. Billings, each showed that such number names are infinite in number. The phrase "added to ten" has ten letters, therefore we can write an infinity of honest names merely by repeating this phrase as many times as desired, e.g., "four added to ten, added to ten, added to ten . . ."

There are many honest numbers, in Dr. Matrix's intended sense, in other languages. Dmitri Borgmann, on page 243 of his marvelous book *Language on Vacation* (New York: Scribner, 1965), points out that the name for 3 in Danish, Norwegian, Swedish, and Italian is TRE, and in Welsh, Irish, and Gaelic it is TRI; 4 is VIER in Dutch and German, FIRE in Danish and Norwegian, and FYRA in Swedish; 5 is CINCO in Spanish and Portuguese, CINCI in Romanian, PIATY in Polish, PENKI in Lithuanian, PEEZI in Lettish, and PENTE in Greek. Peter Salus tells me that it is PAÑCA in Sanskrit.

In *Beyond Language* (New York: Scribner, 1967), pp. 245–246, Borgmann lists honest numbers from O (1 in Middle English) to KUUSTEISTKUMMEND (16 in Estonian). One of the most dishonest numbers in English, he points out in his previous book, is FIVE. Not only does it have four letters, but it also contains at its center IV, the Roman numeral for 4.

Harry Lindgren pointed out in a letter that the consecutive digits 2, 3, 4 have the honest names TO, TRE, FIRE in both Danish and Norwegian, and DU, TRI, KVAR in Esperanto.

For more on honest foreign numbers see David L. Silverman's "Kickshaws" department in *Word Ways* 3 (1970): 46–47; Rudolf Ondrejka's letter in the *Journal of Recreational Mathematics* 4 (1971):151; and Sidney Kravitz, "The Lucky Languages," ibid. 7 (1974):225–228.

Kravitz points out that 4 is also the limit of a sequence of words formed by choosing any English number name, counting its letters, spelling that number, and repeating the process until it converges on 4. Are there other "lucky languages" that converge in this way on a single honest number? Assuming that no number greater than 20 is honest in any language, Kravitz examines seventeen western languages and reports the results. He found only four lucky languages. English, Dutch, and German converge on 4, Italian on 3.

Sin Hitotumatu, whose name means "one pine tree" in Japanese, informed me that if Romanized expressions are used for the names of numbers 1 through 10, there are no honest numbers. The ten words are: *hito* (1), *huta* (2), *mi* (3), *yo* (4), *itu* (5), *mu* (6), *nana* (7), *ya* (8) *koko* (9), and *too* (10). In counting, the Japanese usually add *tu* to each word: *hitotu, hutatu, mittu, yottu,* and so on. An old Japanese riddle asks: "Why does *too* have no *tu*?" Answer: Its brother *itutu* already has two *tus*.

The Japanese, Hitotumatu wrote, also name the first ten numbers by a system that comes from China: *iti, ni, san, si, go, roku, hiti, hati, ku, zyu*. Here the numbers 2 and 3 are both honest. In using this system, the word *si* is often replaced by *yon* to avoid the unhappy fact that *si* is pronounced like the Japanese word for death. Also, *nana* is often substituted for *hiti* to avoid confusion with *iti*. Beyond ten the Japanese system is seldom used, but in this old system, Hitotumatu writes, there is a remarkable honest number *tooaramihuta* (12).

Nine

I. The acrostic sonnet is "An Enigma," by Edgar Allan Poe. The lady's name, Sarah Anna Lewis, is read by taking the first letter of the first line, second letter of the second line, and so on through all fourteen lines. Mrs. Lewis was a Baltimore poet whose efforts were considered "rubbish" by Poe but to whom he was indebted for financial aid. The sonnet is not as well known as his other acrostic, "A Valentine," in which another lady's name is similarly concealed. The word "tuckermanities" in line ten of "An Enigma" refers to the conventional, sentimental work of Henry Theodore Tuckerman, a New York poet and author of the day.

II. The acrostic in the poem by "Maude" is read by taking the first two letters of each line. They spell "Peculiar acrostic."

III. In J. A. Lindon's poem the first line can be read in two other ways: by reading the first word of each line, and the nth word of each nth line.

There is no comprehensive history of acrostic verse. The eleventh edition of the *Encyclopaedia Britannica* has an article on the acrostic, but it seems to be derived chiefly from Bombaugh's book, mentioned in a footnote of chapter 9. On acrostics in the Old Testament and other ancient Hebrew religious writings, consult the *Jewish Encyclopedia,* the *Encyclopedia of Religion and Ethics,* and the *Interpreter's Dictionary of the Bible.* N. K. Gottwald discusses the acrostic form in the first chapter of his book *Studies in the Book of Lamentations* (1954).·

Scores of books of acrostic verse have been published, most of them privately printed. *Dick's Original Album of Verses and Acrostics* (New York: Dick and Fitzgerald, 1879), anonymously edited, contains 218 acrostic poems on the first names of ladies. Lewis C. Scott's *Acrostic Poems and Other Verses,* published by the author in 1924, is a weighty tome of four hundred pages. The author was manager of the Scott Pleating and Button Company in Sioux City, Iowa. His acrostics are mostly on the names of high-school students in various graduating classes. The book includes pictures of the students, pictures of the author and his family, and footnotes explaining all the classical allusions in the poems. Type for the entire book was set by the author.

It is not easy to write poetry more forgettable than Scott's, but the reader will find a prize specimen in W. P. Chilton, Jr.'s *Mansion of the Skies: An Acrostic Poem on the Lord's Prayer* (New York: John Ross and Company, 1875), a book of twenty-seven pages. The poem begins:

O sweet, celestial home—yon gilded sky—
Undimmed in radiance for endless years,
Robed bright in beauty for eternity!
Fain would I sing the bliss which there appears,

An acrostic poem in which the first *words* of each line are the Lord's Prayer will be found on pages 139–140 of Bombaugh's book, cited in chapter 9 above.

Ten

As Malcolm Strasler and other *Scientific American* readers pointed out, if squares are listed consecutively and flush right, it is not only the last column of digits that cyclically repeats a palindromic sequence. *Every* column does, although the sequences get longer as you move left. The second column from the right repeats a palindromic sequence of 43 digits, separated by seven 0's. The third column from the right repeats a palindromic sequence of 481 digits, separated by nineteen 0's. The fourth column from the right's sequence is 4,937 digits separated by sixty-three 0's. From the second column on, the total lengths of the cycles are in the series 50, 500, 5000, ...

I. The formula for numbers whose squares end in 44 is $50x \pm 12$, where x is any nonnegative integer. For numbers whose squares end in 444, the formula is $500x \pm 38$. After 1,444, the next smallest square ending in 444 is $462^2 = 213,444$.

II. To prove that no square of two or more digits can have all odd digits, we show that every square ending in an odd digit must have an even digit second from the end. If a square ends in an odd digit, so must its square root.

Therefore, we have two situations to consider: a square root ending in odd-odd digits and one ending in even-odd digits. The squarings of both numbers are shown schematically in Figure 38, with E for even and O for odd. The dots indicate that each line of numbers may extend any length to the left.

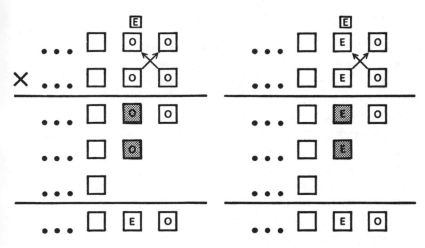

Figure 38. Proof of the odd-digit square problem. Left: square root ends in odd-odd digits. Right: square root ends in even-odd digits

In both cases, when the odd end digits of the square roots are multiplied, the amount carried must be 0, 2, 4, or 8. Since these are all even we place an E inside the two smaller squares to show that the carried amount is even. On the left, each cross multiplication indicated by an arrow gives an odd product that remains odd after the carried amount is added to it. The two shaded digits therefore will be odd, and since the sum of two odd numbers is even, the second digit from the end of the bottom line must be even. The situation on the right is similar, only now the cross multiplications give even products that remain even after the carried amount is added. The two shaded digits therefore will be even. Since their sum is even, the second digit from the end of the bottom line again must be even. Interested readers can go on to prove a stronger theorem, which states that the second digit from the end is odd if and only if the end digit is 6.

III. Number 12 and its reversal have the same curious properties as 13 and 31. Square 12, reverse the digits, extract the square root, and you get 21, the reverse of 12. The digits of either square sum to 9, the square of 3, which is the sum of the digits of 12 or 21.

Frederick S. Parmenter of Troy, New York, read about these numbers in the last chapter of Leonard Eugene Dickson's *History of the Theory of Numbers, Volume I: Divisibility and Primality,* and recently set about the task of searching for all pairs of numbers with the same properties. Let us call such a number a par number. Parmenter proved to his surprise that no par number can contain digits other than 0, 1, 2, 3, and that a number n is par if and only if the sum of its digits is the square root of the sum of the digits of n^2. Since numbers in the series 12, 102, 1002, 10002, . . . , and the series 13, 103, 1003, 10003, . . . , are all pars the set of pars is infinite. If we define a fundamental par as one with no 0's, and include palindromic pars, the set has 55 members, the smallest 1, the largest 111111111. Eliminating the trivial palindromes, the fundamental pars are the following twenty numbers and their reversals: 12, 13, 112, 113, 122, 1112, 1113, 1121, 1122, 1212, 11112, 11113, 11121, 11122, 111112, 111121, 111211, 1111112, 1111121, 1111211. (This had earlier been reported by Father Victor Feser in *Recreational Mathematics Magazine* (Aug. 1962), p. 39.)

IV. The two two-digit automorphic numbers in base-6 arithmetic are 13 and 44.

V. The second hundred-digit base-10 automorph, the companion to the automorph given in Figure 13, is shown in Figure 39. All pairs of automorphic numbers of the same length, in any number system, have a sum of the form 11, 101, 1001, 10001 and so on. In the decimal system, therefore, given one of such a pair, its mate is found by subtracting the last digit from 11 and all other digits from 9. In the base-6 system the last digit is taken from 7, all others from 5.

There is a simple way to expand a base-10 automorph. If it ends in 5, square it, note the nonzero digit closest to the "tail," then prefix this digit (plus any 0's that may follow it) to the tail to make the expanded automorph. For example: $25^2 = 625$. The nonzero digit closest to the tail is 6, so the

6	0	4	6	9	9	2	6	8	0
8	9	1	8	3	0	1	9	7	0
6	1	4	9	0	1	0	9	9	3
7	8	3	3	4	9	0	4	1	9
1	3	6	1	8	8	9	9	9	4
4	2	5	7	6	5	7	6	7	6
9	1	0	3	8	9	0	9	9	5
8	9	3	3	8	0	0	2	2	6
0	7	7	4	3	7	4	0	0	8
1	7	8	7	1	0	9	3	7	6

Figure 39. One hundred digits of the terminal-6
automorphic sequence

expanded automorph is 625. The square of 625 is 390,625. The closest nonzero digit is 9, and since 9 is followed by a 0, we prefix 90 to the tail, 625, to obtain 90,625. The square of 90,625 is 8,212,890,625; therefore, the next automorph is 890,625.

If the automorph ends in 6, the same algorithm is followed except that the closest nonzero digit is subtracted from 10 before it (and any following 0's) are prefixed. For example: $76^2 = 5,776$. The nearest nonzero digit is 7. Take 7 from 10, then prefix 3 to get 376. The two algorithms provide computer programs that will rapidly extend either automorphic series to tens of thousands of digits.

(The base-6 automorphs 13 and 44 can be extended, using similar principles, to seven digits: 1,350,213 and 4,205,344.)

Many readers discovered the two algorithms and wrote to me about them, some sending proofs. Lou Goldman and Ted Katsanis, who together hit on the procedures, also observed that if an n-digit automorph ending in 5 is multiplied by an n-digit automorph ending in 6, the product ends in n 0's. Thus the product of the two hundred-digit automorphs in Figures 13 and 34 is a number that ends in a hundred 0's.

The number of different automorphs for any number system is 2^p, where p is the number of different prime factors in the base. Thus if the base is $2 \times 3 \times 5 = 30$, there are $2^3 = 8$ different automorphs. This includes the trivial cases of 0 and 1, which are automorphic in all number systems.

For more on automorphic numbers, especially in other bases than 10, see:

Vernon deGuerre and R. A. Fairbairn, "Automorphic Numbers," *Journal of Recreational Mathematics* 1 (1968):173–179.

R. A. Fairbairn, "More on Automorphic Numbers," ibid. 2 (1969):170–174.

Gregory Wulczyn, "W-digit Automorphs," *Mathematics Magazine* (March 1969), pp. 99–100.

J. A. H. Hunter coined the term trimorphic for a number whose cube ends with itself. The two automorphic series are, of course, trimorphic. In addition to these, and excluding trivial cases of numbers ending with a series of 9's or a series of 0's followed by 1, Hunter found eight other non-trivial sequences. See his note on "Trimorphic Numbers," *Journal of Recreational Mathematics* 7 (1974):177.

The column reporting my visit to Squaresville appeared in January 1968. On April 7 the *New York Times* reported the appearance in California of a lavishly printed eleven-inch-square magazine called *Square*, "which uses psyche-delic design style and a quasi-hippie approach to cloak its message." The message was one of ultraconservatism. The magazine's main backer was Patrick Frawley, Jr. (note the sixteen letters), a ballpoint pen and razor magnate whose principal business interest is Eversharp (nine letters). The magazine was to appear 2^2 times per year.

Twelve

I. Figure 40 shows one of the four basic solutions (not counting reflections) of the Nixon spelling problem. The *C* and *L*, and the *N* and *X*, can be interchanged. Note the significant *U.S.* at the top.

Figure 40. A solution to the Nixon matrix

The Humphrey matrix has three basic solutions: the one shown in Figure 14, the same with B and Y interchanged, and one completely different:

UM
BEHP
YTRO
AI

The Johnson matrix has eight basic solutions. Many readers sent some of these solutions, with Malcolm C. Holtje and Dean P. McCullough providing all fifteen. Several readers sent graph techniques for enumerating all solutions to such spelling problems.

II. One method of finding the eight different ways sixteen pearl and jade beads can be arranged in a bracelet to show all sixteen possible quadruplet combinations is as follows. Represent the bracelet as a row of binary digits, 1 = jade, 0 = pearl. The chain is considered cyclic, its end joining the front. Combinations 1111 and 0000, which must be on the chain, can be adjacent or separated by 1, 2, 3, or 4 beads. These five possibilities are shown in Figure 41. Since each of the two quadruplets must be bounded on

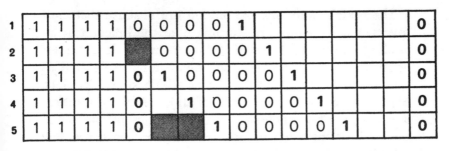

1	1	1	1	1	0	0	0	0	1						0
2	1	1	1	1	▨	0	0	0	0	1					0
3	1	1	1	1	0	1	0	0	0	0	1				0
4	1	1	1	1	0		1	0	0	0	0	1			0
5	1	1	1	1	0	▨	▨	1	0	0	0	0	1		0

Figure 41. Chart for solving bracelet problem

both sides by digits of the other type, we must add the contrasting digits as shown. The second row is immediately eliminated because 1 or 0 in the shaded cell will duplicate 1111 or 0000. Row five is eliminated by trying the four doublets (00, 01, 10, 11) in the two shaded cells and noting that in each case a quadruplet is duplicated.

The remaining rows have solutions. Row one has four, shown below in complementary pairs. Each is obtained from the other by changing all 1's to 0's and all 0's to 1's, which is equivalent to exchanging pearl and jade beads.

$$1111000010011010$$
$$0000111101100101$$

$$1111000010110100$$
$$0000111101001011$$

Row three has one complementary pair:

$$1111010000110010$$
$$0000101111001101$$

Row four has one complementary pair:

$$1111011000010100$$
$$0000100111101011$$

These eight solutions are unique in the sense that their reversals are obtained by turning over the bracelet or by circling it in the opposite direction. A simple algorithm for constructing a minimum-length chain for any desired n-tuplet begins by making a list of all possible n-tuplets. Start the chain with an n-tuplet of all 0's. Add 1 if it makes a new n-tuplet, then check this new n-tuplet off the list. If the 1 fails to make a new n-tuplet, add 0 and cross the new n-tuplet off the list. Continue this procedure to generate a solution.

This algorithm is a special case of a beautiful general procedure, discovered in 1934 by Monroe H. Martin of the University of Maryland, that covers minimum-length bracelets showing all n-tuplets for beads of m different colors. For example, if there are three colors, 0, 1, 2, and we want a bracelet showing all 27 triplets, we start with 000 and proceed to add digits, always selecting the highest digit that will not duplicate a triplet that has already been formed. The result is 00022212202112102012001101. The procedure is given in exercise 17, page 33, of the second volume of Donald E. Knuth's monumental series, *The Art of Computer Programming* (Reading, Mass.: Addison-Wesley, 1969). The book is as rich in recreational material and little-known historical sidelights as the first volume, and I recommend it highly.

In volume 1 (answer to exercise 23, p. 379) Knuth gives a remarkable formula (due to N. G. de Bruijn of Holland) that provides the number of minimum-length bracelets of n-tuplets and m colors, including reversals as different. Knuth tells me that when reversals are *not* considered different (as in the problem given here), the formula is

$$\frac{\frac{1}{2}(m\,!)^{m^{n-1}}}{m^n}$$

If reversals are considered different, the formula is simply doubled. It is not hard to prove, Knuth adds, that no bracelets are symmetrical in the sense that they are identical with their reversals. The only exception is the four-bead doublet bracelet of two colors; in its case the above formula gives ½ as the number of different bracelets instead of 1, the correct number.

Different methods, including graph techniques, for constructing such bracelets are discussed in chapter 10 of Sherman K. Stein's *Mathematics: The Man-Made Uni-*

verse, 2nd ed. (San Francisco: W. H. Freeman and Company, 1969). The chapter is an expansion of Stein's article in *Scientific American* (May 1961). It includes a fascinating history of the problem, beginning with the use of such chains by poets in India a thousand years ago (as mnemonic devices for remembering combinations of long and short beats) and ending with current applications, particularly in the construction of what communication theorists call error-correcting codes. There is a good list of references.

H. M. Schweighofer wrote to describe how these binary chains are "widely used in remote-control devices such as the remote tuning of airline and military radios. Instead of using different beads, of course, the code is formed by contacting respectively the front (1) or rear (0) rotor blades of a wafer switch. The control circuit used is a 'reentrant' circuit whereby a current flows whenever the remote-control and motor-driven seeking switches are not in the same position (same code). The use of wafer switches to create the code sequences in a manner functionally identical to your bead bracelets was, in fact, one feature of my first patent.

"Your article brought back some fond memories of the days I spent, some twenty years ago, designing switching sequences based on these principles. Now that radio tuning has also gone solid-state there is less need for rotary switches and such sequential binary codes. Standard binary or BCD codes are used in the new designs. However, many of your readers have undoubtedly used equipment operated on the principles of Iva's bead bracelet."

A more esoteric application of bracelet codes is to magic tricks. For readers interested in mathematical magic, and with access to the literature, here are a few references. The first use of the principle seems to be in a trick called

"Coluria," in Charles T. Jordan's *Thirty Card Mysteries* (1919). The principle also underlies "Suitability," in *Card Mysteries* by William Larsen and T. Page Wright, and a Bible book test by Robert Hummer in his *Six Tricks for 1944*. The magic periodical *Pallbearers Review* has published several card tricks utilizing bracelet codes. See, for example, "Other Voices II," by Ben Christopher (Aug. 1968), and my "Yin-Yang" card trick (Feb. 1974).

Thirteen

I. The anagram of MOON STARERS is ASTRONOMERS.

II. The solution to SPIRO × 7 = AGNEW is 14,076 × 7 = 98,532. This unique solution was first published by E. P. Starke in the *American Mathematical Monthly*, 53 (1946): 4–5. In *Mathematics Magazine* 42 (1969):102–103, David Daykin gives two tables of computer results that list the number of different solutions of $A = kB$, in all number systems with bases 3 through 15, where k is any number from 2 through 14 and A and B are numbers that together contain all ten digits once each or all nine digits (0 excluded) once each. Solutions in which A or B begin with 0 are not considered. In Dr. Matrix's problem (base 10 and all ten digits) there are forty-eight solutions for $k = 2$, six for $k = 3$, eight for $k = 4$, twelve for $k = 5$, none for $k = 6$, one for $k = 7$, sixteen for $k = 8$, and three for $k = 9$.

M. M. Williams, senior engineer at the Kennedy Space Center, sent the following comments on this cryptarithm: "It should be noted that 1407698532 can be viewed as 14/07/69, 14 July 1969, which was T − 2 days in the *Apollo 11* countdown, and 8/5/32, August 5, 1932, which was astronaut Armstrong's second birthday. The numbers 0932, the time of lift-off, also appear in sequence. The sum of 9 plus 2 is 11, 3 was the number of astronauts, and 2 was the number of men to walk on the moon. The last three digits of the product, 532, are the time (EDT) at Cape Kennedy when it was 0932 hours GMT."

III. Years that have the property of appearing inverted at the end of their squares must begin with 1 and end in 9. Aside from 1969 and the trivial case of 1, the only such year before the year 10000 is 19.

IV. Readers interested in learning more about the problem of placing n points on a sphere should consult:

H. S. M. Coxeter, "The Problem of Packing a Number of Equal Nonoverlapping Circles on a Sphere," *Transactions of the New York Academy of Sciences*, series II, 24 (1962): 320–331.

L. Fejes Toth, *Regular Figures* (Elmsford, N.Y.: Pergamon, 1964), chapter 7.

Michael Goldberg, "Axially Symmetric Packing of Equal Circles on a Sphere," *Annales Universitatis Scientiarum* (Budapest) 10 (1967).

No general solution to the problem is within sight.

V. A year can have no more than four perverse months and no fewer than two. It can have no more than three Friday the 13ths and no fewer than one. The number of perverse months in a year plus the number of Friday the 13ths is four—except for nonleap years beginning on Sunday or Thursday, and leap years beginning on Sunday or Saturday, when the sum is five. The last year with four perverse months was 1972, with no more to come in this century.

A proof by B. H. Brown that the 13th is more likely to fall on Friday than on any other day appeared in the *Amer-*

ican Mathematical Monthly 40 (1933):607; a proof by
S. R. Baxter (done at the age of thirteen) appeared in the
Mathematical Gazette 53 (1969):127–129.

Many short proofs of the maximum and minimum
numbers of Friday the 13ths in a year have been published. For recent discussions of this problem see William
T. Bailey, "Friday-the-Thirteenth," *Mathematics Teacher*
62 (1969):363–364; J. O. Irwin, "Friday 13th," *Mathematical Gazette* 55 (1971):412–415; and John Wagner and
Robert McGinty, "Superstitious?" *Mathematics Teacher*
65 (1972):503–505.

Fourteen

I. The other number smaller than 10,000 that has sixty-three proper divisors (including 1 but not the number itself) is 7,560.

Panos D. Bardis, in "Overpopulation, the Ideal City, and Plato's Mathematics," *Platon* 23 (1971):129–131, suggests that Plato may have chosen the number 5,040 because it is the factorial of the mystic number 7. When I mentioned this to Dr. Matrix he immediately replied, "Yes, and if Plato visited the city its population would become the square of 71."

II. Here is the only way to arrange two three-digit square numbers in a two-by-three matrix so that each column, read from the top down, is a two-digit square:

$$841$$
$$196$$

I am indebted to letters from Mrs. Joseph Charles, S. S. Chebli, Joel R. Cohen, Edwin M. McMillan, H. Martin Pitts, Emile C. Van Remoortere, and Richard S. Watt for suggestions about the name of Martin Luther King, Jr.; and to Sally Porter Jenks, Thomas R. Jones, and Eugene McGovern for names related to a profession.

Fifteen

I. The year with the largest time span between itself and its inverse is 1066, the date of the Norman Conquest. The span is 9901 − 1066 = 8,835 years. This assumes that B.C. dates are excluded, as well as initial 0's of numbers: e.g., 6 A.D., turned upside down, is 9, not 9000.

II. The only state capital that does not share a letter with its state is Pierre, South Dakota.

III. The only letter not in the name of any state is q.

IV. The only heteroliteral day-month pair, aside from Friday-June, is Sunday-October.

V. The only letter not in number names from 0 through 99, but in the name of every number from 100 through 999,999, is d.

The NY to OZ word shift was discovered by Mary Scott, a member of the International Wizard of Oz Club. The OZ to PA shift was first noted by Elliott Weiss, of Philadelphia, who did not know its significance when he mentioned it in a letter to Dr. Matrix. Dr. Matrix, a lifelong Oz buff, recognized the synchronicity at once.

Sixteen

The California school was indeed founded by Dr. Matrix. It flourished until November 1973, when the Amazing Randi, having enrolled at the school under his real name, James Zwinge, wrote a detailed exposé of the school's deceptive practices for the *Los Angeles Times.* The school closed shortly thereafter.

As most readers of my column perceived, all but ten of the items, if instructions are carefully followed, can have only one result. No one, therefore, can score less than sixteen. Moreover, each of the uncertain tests give the reader a high probability of a hit.

Dr. Matrix was kind enough, on a later occasion, to reveal his sources for all twenty-six items:

1. Martin Gardner, *Scientific American Book of Mathematical Puzzles and Diversions,* chapter 2.

2. Invented by magician Frederic DeMuth.

3. A card trick called "Miraskill," devised by Stewart James.

4. An old "psychological force" that was much used by the Israeli magician Uri Geller. Uri continued to use it in spite of its appearance in my 1973 column. Charles Panati, of *Newsweek's* science staff, was enormously impressed when Uri performed this successfully on him during their

first meeting in 1973. As late as 1975 Uri was still doing it in public appearances.

If a person puts a triangle inside a circle—as happened when Uri appeared on Dick Klinger's morning TV talk show on KGW in Portland, Oregon, June 17, 1975—Uri naturally calls it a hit. In fact, said Uri to Klinger, it was a *remarkable* hit "because most people usually draw a square or a cross." This, of course, is not true. Most people draw a circle and a triangle. Why is it so successful? Ask someone not a mathematician to name a few simple geometrical figures. It will be unusual if he can think of more than four.

5. An old magic trick. The principle is the same as in item 4.

6. Devised by Victor Eigen, the mathemagician interviewed in chapter 9 of Martin Gardner's *New Mathematical Diversions from Scientific American* (New York: Simon and Schuster, 1966). The probability of getting AND is 25/36.

7. Another old trick.

8. A numerical psychological force of unknown origin.

9. The invention of magician Robert Hummer.

10. An old mathematical trick.

11. Another oldie. It is given by L. N. Chapin, *The Beautiful, the Wonderful, and the Wise,* (Chicago: John C. Winston, 1885), p. 458, but is surely older.

12. Devised by Fitch Cheney, mathematician and magician.

13. From Victor Eigen. For variations and presentation, see my "Paperfold Prediction," *Swami* (a magic periodical published in Calcutta), July 1973.

14. An old digital-root trick.

15. From Victor Eigen.

16. Another psychological force popularized by Uri Gel-

ler. London and Paris are the two cities most often selected. Uri often writes London, crosses it out, then writes
Paris. If Paris is correct he explains that London was his
first impression but he quickly realized it was wrong. If
London is correct he says, "I should have trusted my first
impression, but I was confused when I got a stronger impression of Paris. Did anyone else here think of Paris?"
Several people usually say yes, so Uri wins either way.

17. Based on a trick by British magician Jack Yates.

18. An old favorite of magician Dai Vernon.

19. A trick by Walter B. Gibson, writer and magician.

20. Another psychological force.

21. A card trick by magician Henry Christ.

22. Another psychological number force, usually paired
with item 8.

23. From Victor Eigen.

24. An old psychological force.

25. A trick by Robert Hummer.

26. From Victor Eigen. The probability of getting THE
is 4/5.

The tricks are excellent stunts for a teacher to show a
mathematics class. Having students explain why the mathematical ones work—often with the aid of simple equations—can be a stimulating classroom project.

Seventeen

The pyramid labeling problem was original with Dr. Matrix. It can be attacked by writing a set of Diophantine equations, but here is an informal way of getting the same result.

Five vertexes, each summing to 16, make a total of $5 \times 16 = 80$. Because each edge number contributes to two vertexes, the sum of the eight numbers must be $80/2 = 40$. Only three sets of eight different positive integers, each no greater than 10, add to 40. They are:

1, 2, 3, 4, 5, 7, 8, 10
1, 2, 3, 4, 6, 7, 8, 9
1, 2, 3, 4, 5, 6, 9, 10

Consider the first set. The number 10 must label either a base edge or an edge on the pyramid's side. If it is a base edge, then at each end it must meet a pair of edges with a sum of 6. The only possible doublets are 1, 5 and 4, 2. By trying them in their four possible arrangements, it is easy to see that there is no way to complete the labeling to make the pyramid magic. If 10 labels a side edge, the other three edges meeting at the apex must add to 6. The only possible triplet is 1, 2, 3. Considering the symmetry, there are three arrangements (the number opposite 10 can be 1, 2, or 3). None permits completion of the labeling.

Consider the second set. If 9 is a base edge, the pairs meeting at the two ends must be 6, 1 and 4, 3. Trying them in their four arrangements does not lead to a solution. If 9 is a side edge, the other three edges meeting at the apex must be 1, 2, 4, in one of three arrangements. Again, no solution is possible.

Consider the third set. If 10 is a side edge, the triplet meeting 10 at the top must be 1, 2, 3. It permits no solution. If 10 is a base edge, the doublets at the ends must be 1, 5 and 4, 2. This allows only the solution shown in Figure 42. It is unique except for rotations and reflections.

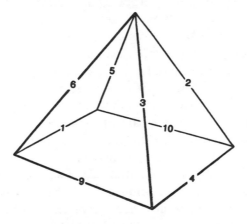

Figure 42. The magic pyramid

Hiram Fuller Gutgasz wrote to point out that the solution is unique in a stronger sense than Dr. Matrix realized. It is unnecessary to restrict the labeling numbers to positive integers no greater than 10. The solution is unique, when the magic constant is 16, for *any* set of eight distinct positive integers. If the labeling is confined to the ten digits (0 through 9), there are two solutions. Both omit 1 and 4. Using numbers from 1 through 10, only one other magic pyramid is possible. It is a pyramid with a constant of 18, and a unique labeling that omits 1 and 9.

Two readers, J. A. McCallum and Sheldon B. Akers, each asked themselves if a magic pyramid could be constructed with eight consecutive positive integers. Each found the only solution and proved it unique. The numbers are 4, 5, 6, 7, 8, 9, 10, 11, with a magic constant of 24.

Who would have thought it possible that anyone who read chapter 17 when it appeared in *Scientific American* (June 1974) could have taken Dr. Matrix's research seriously? Yet so pervasive is the public's current obsession with all things psychic that hundreds of readers asked for additional information about psi-org energy and how they could get in touch with Dr. Matrix.

Many readers drove to Pyramid Lake looking for the factory, some steering their cars up dangerous canyon roads. A professor from the California Polytechnic State University, who failed to locate the factory, wrote me in great annoyance. He informed me that the town of Pyramid consists of only one farmhouse and several outhouses. Inhabitants told him they had never heard of Dr. Matrix. Nor had anyone on the reservation, to whom he spoke, heard of One-Tooth Ree. Another reader, in Pukalani, Hawaii, said my column had "really turned people on" in his area. One lady had moved into a glass and plastic pyramid beside a jungle stream, and a health-food store was keeping its produce under a giant pyramid. He and his friends wanted to pay my fare to Hawaii so I could lecture to them about the new discoveries.

The most startling letter came from a prominent, well-known New York City publisher. He offered me an advance of $15,000 for a book, which I was to write quickly, titled *Pyramid Power*. He was convinced he could promote it into a "top best-seller." I had lunch with him to see whether he was serious. He was. When he learned that nothing Dr. Matrix said about psi-org energy was authentic, he proposed that I write the book under a pseudonym. He was even willing to let me expose the book later as a hoax provided he had a full year to promote it as a serious study. He thought it would add to the book's sales if I could tie the pyramid in some way to messages from outer space, perhaps by attaching an aerial to the pyramid.

The book he proposed might well have been promoted into another *Secret Life of Plants*. As I type this postscript, in the fall of 1975, the pyramid cult shows no sign of abating in spite of the fact that leading parapsychologists— Thelma Moss, for instance—have declared the pyramids to be worthless.

New Horizons, published by the Toronto Society for Psychic Research, ran three articles on pyramid power in its Summer 1973 issue. Allan Alter reported on extensive tests showing that pyramid containers "are no more effective than those of other shapes in preserving and dehydrating organic material." (He allowed, though, that the Great Pyramid of Egypt may contain a "mysterious force" because of the geomagnetic field at its site.) Dale Simmons reported on tests with razor blades, complete with microphotos, that failed to show any influence of the pyramid on the blades.

Such negative opinions seldom reach the gullible, and have no effect even when they do. Edmund Scientific Company continues to advertise its transparent plastic model of the pyramid at the outrageous markup price of $20. The ads appeared regularly in *Science News*, where they were read by thousands of high-school students.

A friend in Washington, D.C., Cornelius van S. Roosevelt, wrote me in 1975 that he had been a judge of high-school exhibits at an annual science fair. Two exhibits were on pyramid power. "At first I thought these were structured to disprove the fallacy," he wrote, "but I was mistaken. Each claimed to have proved that razor blades were in fact sharpened. . . ." Both exhibits, done independently at different schools, cited my column as a major reference.

Models of the pyramid, in all sizes and materials, continue to be advertised in occult magazines. Flanagan's Pyramid Products, Glendale, California, in addition to tents sells a PEG or "pyramid energy generator" (fifteen little pyramids side by side), pyramid energy plates, and something called the Flanagan Spiral Generator. This is a gold spiral that creates "bioenergy similar to but much stronger than the Pyramid." Also on sale are two books by Flanagan: *Pyramid Power* (hardcover) and *The Pyramid in Its Relation to Biocosmic Energy* (paperback). Max Toth and Greg Nielsen have also written a book called *Pyramid Power*, published in soft covers in 1974 by Freeway Press.

Flanagan, who seems to be the world's number one promoter of pyramid power, likes to put a Ph.D. after his name, but I have been unable to discover when or where he received it. The introduction to one of his books says he has been listed in *Who's Who in American Science* since 1962, but there is no volume with this title nor is he listed in any editions of *American Men of Science*.

Life magazine published a special issue (Sept. 14, 1962) devoted to "The Take-over Generation: One Hundred of the Most Important Young Men and Women in the United States." It contains two pages about Pat, who at the time was seventeen and living in Bellaire, Texas. William Moeser, author of the article ("Whiz Kid, Hands Down"), describes Pat as the "model of a mature and inquisitive scientist."

Why was Pat included among the hundred? He had invented a device called the "neurophone," said to transmit electrical pulses, produced by sound, directly to the body's nervous system and thence to the brain. You can plug up your ears, place the machine's output "muffs" to your temples, and hear. No one, not even Pat, says the author, knows why it works, but several companies are "interested." It is suggested that a similar device may be able to do the same thing with light, thus enabling the blind to see. No details about this marvelous invention are given, so it is impossible to know if a new principle is involved or whether it simply transmits vibrations to the skull, where they vibrate the inner ear like those patented devices of the nineteenth century for hearing through the teeth.

There is a photograph of young Pat standing on his head, and another showing him grinning while two teenage girls listen with neurophone muffs to their temples. A picture of a more mature Pat appears in Tom Valentine's paperback, *The Great Pyramid* (New York: Pinnacle, 1975), and on the dust jacket of the second edition (1975) of Pat's *Pyramid Power*. Another good account of Dr. Flanagan's revolutionary research is given in *Exploring the Great Pyramid Shape in America*, by Kim Russell, published by Shapeous Researching, Dallas, Tex.

The Parker Publishing Company, West Nyack, New York, continues to take full-page ads in *Fate* to sell Al G. Manning's book *The Miracle of Universal Psychic Power: How to Pyramid Your Way to Prosperity*. This valuable scientific work costs a mere $7.95. Almost immediately after using a pyramid, says the ad, "D. J. reports finding ten crisp $100 bills in her kitchen drawer!" The Pyramid Think Tank Research Company, in Ingram, Texas, sells a model that comes with a magnetic compass so you can align it properly. "Some users report a tremendous increase in sensitivity, especially in the field of ESP and psychic powers. Others indicate dramatic improvement in overcoming depression, sharpening of consciousness, and amazing rejuvenating effects." The model costs only $19.95, five cents less than the Edmund model.

Science Digest (March 1975, p. 91) carried an ad for a "Pyramid Orgone Tent." This plastic tent comes "ready to erect" and a "compass square and meditation program" are included. The largest tent costs $40. The ad was placed by one J. Kennedy, of Biloxi, Mississippi. Sensonics, Inc., Maitland, Florida, sells the largest pyramid tent of all—forty feet high—and issues a periodical called *Boogie-on* that reports on their latest research developments.

The leading periodical on pyramid power is the *Pyramid Guide,* a bimonthly that started in September 1972. It is edited and published by Bill Cox and Georgiana Teeple, who run El Cariso Publications, Elsinore, California. The guide's pages relate pyramid power to Edgar Cayce, dowsing, time warps, orgone, the Hieronymus machine, psychometry, levitation, meditation, biorhythm, the Keely motor, Findhorn, color healing, theosophy, auras, Kirlian photography, and of course all aspects of ESP.

Pyramid models have been found by El Cariso to improve TV reception, recharge flashlight batteries, and cause one's hand to levitate. They will also sharpen *electric* razors. The firm sells pipes for making a "panelless" tent consisting of just the eight edges of a pyramid. The power "develops along ridges and corners without loss of energy." The energy is the same as Baron von Reichenbach's "odic force" and Wilhelm Reich's "orgone." One reader reports getting over one hundred shaves from one Gillette Blue Blade by keeping it in his orgone box. The box, he adds, has a great advantage over the pyramid because it needn't be oriented north.

El Cariso also sells a metal cone that apparently works as well as the pyramid, an "aurameter" for dowsing, and bumper stickers that read: "There is no need to conserve Pyramid energy. The source is ETERNAL—the supply INFINITE." The guide reports that the Reverend John D. Rankin, pastor of Unity Church, in Houston, has built a Pyramid Church of anodized aluminum in which he and his congregation meditate.

In Canada pyramid power soared in 1976. According to Wayne Lilley, writing on "The Pyramid Pushers" in Canada's *Financial Post Magazine,* April, 1976, pages 19–24, half a dozen companies in the Toronto area were manufacturing pyramid models, in all sizes, to meet the growing demand. Manufacturers maintained that the University of Guelph, a respected Canadian agricultural school, had verified the influence of the pyramids on plant growth, but when Lilley checked, Guelph professors vigorously denied it. On the contrary, they told him, their tests showed that the pyramids had no effect whatever on plant growth. Rumors that NASA, in the U.S., was sponsoring pyramid research were denied by NASA.

One of the most enthusiastic promoters of Canadian pyramid power is Red Kelly, coach of the Toronto hockey team, the Maple Leafs. After his daughter found that a pyramid under her pillow stopped her migraine headaches, Coach Kelly began hanging pyramids on the ceiling of his team's dressing room and putting them under benches. See the Toronto *Globe and Mail,* April 24, for a photograph of Red Kelly with a pyramid on his head, and "Pyramid Power: Extra Energy Flows to Leaf Players," on the sports page of the Toronto *Star,* April 24.

One might suppose that articles on pyramid power run only in crummy magazines like *Fate,* or flimsy little publications such as *Pyramid Guide.* Not so. The *New York Times* (April 20, 1974) devoted half its family page to a story about Michael Reynolds, a young architect in Taos, New Mexico. Reynolds sleeps at night inside a large model of the pyramid. First he experienced flashes of precognition about friends coming to see him, then it was a "high-pitched sound" that he heard only in early morning hours.

He had strange color dreams. "They are as vivid," he said, "as the pictures on a poorly tuned television set. I suppose that the Cheops Pyramid must be like a finely tuned television set." The pyramid has increased his vitality and reduced his need for sleep.

Postscript (1985)

High praise for Flanagan's neurophone surfaced in *Analog Science Fiction* in two "Alternate View" columns by G. Harry Stine (July 1979 and February 1980). The second column is about Stine's visit to Flanagan's ranch and laboratory in Tucson, Arizona, where he tested an improved neurophone. Convinced that the device is no "humbug," Stine wrote: "I don't give a doodly-damn what all you bloody experts out there have presumptuously stated in your outraged fan letters to me." He has put these letters, he adds, in his Utter Bilge file, the name deriving from a famous statement by Sir Richard van der Riet Wooley when he became British Astronomer Royal in 1956. "Space travel," declared the astronomer, "is utter bilge."

Anyone interested in the neurophone can write to the U.S. Patent Office for its two patents: 3,393,279 (July 16, 1968) and 3,647,970 (March 7, 1972). In 1984 Pat was living at 7 Commercial Blvd., Novato, CA 94947. You can obtain from him the latest issue of his Flanagan Research Report, a new book called *Pyramid Power II*, and a catalog of his latest devices for increasing your psychic powers, rejuvenating your sex, and reversing your aging.

Eighteen

I. Alan Wayne's cryptarithm, assuming the usual convention that numbers beginning with 0 are not allowed, has the unique solution:

$$942 + 942 + 942 = 1413 + 1413$$

The digits of 1,413 are the first four digits of pi backward, and $942/3 = 314$, the first three digits of pi.

If numbers starting with 0 are permitted, there is one other solution: $472 + 472 + 472 = 0708 + 0708$. Terry Terman and Daniel S. Marcus have each noted that $472 \times 3 = 1,416$, thus providing the first five digits of pi (the decimal point replaced by the equal sign, and the last digit rounded). Herb Freedman added that $472 = 314 + 158$, the six numbers on the right differing from the first six digits of pi only in the last digit being 8 instead of 9.

II. The thirty-six forms of the hollow square of 144,000 saints are found by letting a be the side of the square array, b the side of the interior square hole, then finding all solutions in positive integers for the Diophantine equation: $a^2 - b^2 = 144,000$. The smallest value for a is 380, making the hollow a square of 20 on the side.

Webb Simmons sent an explanation for the perfect

square in Mrs. White's vision that does not require a hollow. Perhaps the saints were in a trapezoidal formation of 375 ranks, with 197 in the front row, 198 in the next, and so on to the final row of 571. If Mrs. White viewed this trapezoid from an elevated position, perspective would make it appear to her as a perfect solid square.

III. The request for four different positive integers that can be arranged to make a complex fraction $(a/b)/(c/d)$ that equals $(d/c)/(b/a)$ was intended as a joke. The two expressions are easily shown to be equivalent when any real numbers whatever are plugged into the formula.

Many other number jokes are scattered through Dr. Matrix's whimsical commentary. When he mentions, for example, that 491 (the number of the unforgivable sin) is the difference between the squares of consecutive integers, he does not tell his readers that *any* odd number is the difference between the squares of two consecutive integers.

IV. The answer to the first enigma is DAVID; to the second, Lot's wife. The rebus is "No a; h!" or Noah. Dmitri Borgmann informs me that there are many similar single-letter biblical rebuses: b = "Aha—b!" (Ahab), m = "Ha—m!" (Ham), t = "Lo—t!" (Lot), and so on.

Nineteen

I. How did Dr. Matrix vanish the goblet of wine? The goblet was made of ice and was kept in its mold inside a freezer until Dr. Matrix was ready to perform. At the center of the table a small hole at the base of a slightly conical metal disk led into a hollow central leg. The disk was wired so that it became hot when Dr. Matrix pushed a concealed button on his desk. When the goblet melted, the water and the wine drained into the table leg.

II. Here is a simple proof that the recursive formula $(b + 1)/a$ generates a sequence of numbers that loops with period 5. Let a be the first number of the sequence and b the second. When the formula is applied recursively, the third number is $(b + 1)/a$, the fourth is $(a + b + 1)/ab$, the fifth is $(a + 1)/b$, the sixth is a and the seventh is b.

Sam Dalal, the Calcutta magician, is a real person. I obtained his permission to put him into the column, and the resulting publicity brought him many new subscriptions to *Mantra*. After the column appeared, Sam called my attention to a method of vanishing a goblet of wine that is similar to the ice method I described. In *Dunninger's Complete Encyclopedia of Magic* it is suggested that the glass be made of wax that melts on the hot plate. The ice method I described had been told to me by the famous magician Okito, whom I got to

My Calcutta friend Sam Dalal

know after he retired in Chicago. He said he had seen it performed in the home of a European amateur magician, apparently the trick's inventor.

Here is an equally curious method of producing a goblet of
wine. The goblet is made of glass with the same refractive index
as water. Under water, such a glass is completely invisible. The
magician displays a transparent glass bowl that seems to contain
nothing but water. He covers it with a large cloth, lifts out the
hidden goblet, filled with water. A small pellet of red coloring

matter is dropped into the goblet before the magician removes the cloth.

I forgot to mention in the column that the nonest giggle is produced by repeating *e*, the fifth letter of the alphabet, five times. Note that *e* is a famous transcendental number.

Charles Demuth's painting of the numeral 5, the original of which hangs in the Metropolitan Museum of Art (see Figure 12, p. 80), was inspired by the following short poem by William Carlos Williams:

> Among the rain
> and lights
> I saw the figure 5
> in gold
> on a red
> firetruck
> moving
> tense
> unheeded
> to gong clangs
> siren howls
> and wheels rumbling
> through the dark city.

Note that the poem, excluding the numeral, has 30 words. Thirty is the product of 5 and 6, and the poem has just five words of six letters each. For more on 5 see "The Prevalence of Fives," by Gerald Oster, *Natural History*, March 1975, and "The Ubiquitous Number Five," by I. A. Barnett, *Mathematics teacher*, April 1968. On the theorem that the number of divisions needed to find the greatest common denominator of two numbers is five times the number of digits in the smaller number, see Howard Grossman's elegant proof in the *American Mathematical Monthly*, Vol. 31, 1924, page 443. The proof is also given in W. Sierpinski, *Theory of*

Numbers (1964), pages 21-22, and Ross Honsberger, *Mathematical Gems II* (1976), Chapter 7.

Some Indian ragas actually do give the impression of a steadily increasing beat that never gets faster. The precise techniques for creating this illusion were worked out by Ken Knowlton, of Bell Labs. The American composer Elliot Carter used the principle in some of his musical compositions. The illusion is a rhythmic analog of an endlessly rising tone illusion discovered by Roger N. Shepard when he worked at Bell Labs. Shepard's "endless octave," as it has been called, used discrete notes, but Jean-Claude Risset found a way to do it with a continuously rising (or falling) glissando.

Graham Holmes, a Clarkson College engineer, wrote to inform me that if Dr. Matrix were capable of moving one hand fast enough, and stopping it quickly enough, it would produce the sound of a clap. This, in fact, is how the snapping sound of a whip is produced.

Quentin G. Furlow called my attention to a remarkable fact about 5 that occurs in the geometry of higher spaces. A one-dimensional "sphere" of unit diameter (it is a straight line segment) will fit inside a two-dimensional unit "sphere" (circle), and the unit circle will fit inside a unit ball in 3-space. The ball will go inside a unit hypersphere in 4-space, and this in turn fits into a 5-space unit hypersphere. But at this point an incredible turn occurs. The 6-space hypersphere, and all hyperspheres of higher order, will each go into the 5-space sphere! In other words, the 5-space ball is, in a sense, the largest possible unit hypersphere.

Readers familiar with Transcendental Meditation (TM), est, and Scientology will recognize many word plays on terms related to the three cults. My Bagel Lox of Dr. Pepper is a play on Donald Cox, a former vice-president of Coca Cola who became president of est. The Nonestic Ocean is the ocean Dorothy and her yellow hen cross on a raft in *Ozma of Oz*.

Dr. Matrix's procedure of starting with two numbers *a* and *b*, and recursively applying the function $(b + 1)/a$ (where a and b refer to the previous two terms of the series as it is generated by the formula), leads into a vast area of research where there are many unsolved problems. In general, one seeks a rational function that can be applied recursively to n variables, and which will loop in a maximum number of steps. The general problem, which has applications in telephone communication technology, is discussed by two Bell Labs mathematicians, Robert P. Kurshan and Bhaskarpillai Gopinath in "Recursively Generated Periodic Sequences," *Canadian Journal of Mathematics*, Vol. 26 (1974), pages 1356-71, and by J. H. Conway and R. L. Graham in "On Periodic Sequences Defined by Recurrences," an unpublished Bell Labs report.

The formula for two variables that loops in five steps was published by R. C. Lyness in the *Mathematical Gazette*, Vol. 26 (1942), page 62 (see also the same journal, Vol. 29, 1945, page 251 for another note by Lyness). It is not known whether there are rational recursive functions for two variables that loop in longer cycles.

It is not necessary to restrict the two variables to positive real numbers. Either or both numbers may be negative (provided zero is excluded or any step in which there is division by zero) or even complex numbers.

For one variable the simplest formula of this type is $1/a$, which loops with period 2. A maximum period of 3 is obtained by $1/(1-a)$. Just start with any nonzero number, real or complex, take it from 1, find the reciprocal, then repeat two more times. You are back at the starting number.

For three variables, the longest known cycle is 8. It is given by the recurrence function $(1 + b + c)/a$. The letters stand, of course, for the last three terms of the sequence as it is recursively generated. Thus if $a = 1$, $b = 2$, $c = 3$, the sequence is 1, 2, 3, 6, 5, 4, 5/3, 4/3, 1, 2, 3, ... The longest known cycle for four variables is 12. It is produced by $ad/(ac - b)$, a function discovered by Conway.

Andrew Lenard, of Indiana University, sent an interesting letter in which he showed that the five-step formula for two variables is equivalent to a relationship involving points on the real projective plane and the cyclic symmetries of their cross-ratios. When he passed this along to the Canadian geometer H. M. S. Coxeter, he learned from Coxeter that he had rediscovered a 5-cycle theorem of Gauss, and its projective proof.

In ancient Hinduism, Shiva is one of the three manifestations of Brahman, the ultimate incomprehensible ground of being. The other two manifestations are Vishnu and Brahma. While Vishnu sleeps on the cobra shesha, a lotus grows from Vishnu's navel. From a lotus bud emerges Brahma who creates the universe. When the universe is destroyed by Shiva, it is absorbed into Vishnu who sleeps for a "night of Brahma," equivalent to 4,320,000,000 of our years. This process is repeated, the new universe lasting for a "day of Brahma," the same length as the "night." A "year" of Brahma is 360 Brahma "days." After 100 "years" of Brahma, Vishnu is absorbed back into Brahman and for 100 "years" there is nothing but Brahman. Then a new Vishnu will appear.

We are now in the fiftieth year of our Vishnu. For a surprising defense of the view that nothing exists in any strong sense because the universe is a dream, see the closing pages of Mark Twain's *Mysterious Stranger*.

Twenty

I. The problem is to prove that no number starting with 9 and followed by digits in cyclic descending consecutive order (with or without 0) can be prime. All primes except 2 must end in 1, 3, 7, or 9 because if the last digit is even, the number is divisible by 2, and if the last digit is 5, the number is divisible by 5. It is easily shown, by adding digits and reducing to the digital root, that if the sequence terminates with 1, 3, 7, or 9, the digital root must be a multiple of 3, proving that the number also is a multiple of 3.

II. In an alphabetical list of the English spellings for the integers 0 through 1,000 the next to last entry is two hundred two.

III. In an alphabetical list of the Roman numerals from 1 through 1,000 the last entry is XXXVIII, or 38.

IV. The smallest number whose name contains all five vowels plus y is one thousand twenty-five.

V. The next term in the sequence 10^3, 10^9, 10^{27}, 10^2, 10^0 . . . is $10^{.60206}$ . . ., or 4. The first term, "one thousand," is the smallest positive integer whose English name contains a; the second term, "one billion," is the smallest to contain b;

the third, "one octillion," is the smallest to contain c; the fourth, "one hundred," is the smallest to contain d; the fifth, "one," is the smallest to contain e, and the sixth, "four," is the smallest to contain f.

VI. The "unfolded" cubes shown in Figure 43 demonstrate how three cubes can be given lower-case letters so that the cubes can be arranged in a row to spell the first three letters of any month. Note that this is possible because u and n and p and d are inverses of each other.

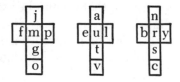

Figure 43. Solution to the calendar cubes problem

Readers pointed out two other inversions that could be exploited: when b is inverted it looks like a g, and when a is turned upside down it looks like an e. Moreover, when u is given a quarter-turn it becomes c. With these new freedoms, one can solve the problem by using only five faces of each die, leaving three faces for pictures. Here is one way to do it:

Die 1	Die 2	Die 3
j	a, e	u, n, c
m	u, n, c	f
p, d	v	r
g, b	l	y
o	t	s

In commenting on problem 3, Uzi Ritte, an Israeli geneticist, informed me that if the Hebrew names for numbers 1 through 999 are alphabetized, the first item is the Hebrew word for 1, and the last is the Hebrew word for 999.

David Emmanuel wrote to contest Dr. Matrix's claim that numerological theorems are as numerous as the primes. They are *more* numerous, he argued. The primes are only countably infinite (they can be put into one-to-one correspondence with the counting numbers) whereas numerological theorems are surely as uncountably infinite as real numbers.

The following articles, in the *Journal of Recreational Mathematics*, deal with primes that have their digits in ascending or descending cyclic order:

"A Consecutive-Digit Prime," Joseph Madachy, Vol. 4, April 1971, page 100.

"Consecutive-Digit Primes — Again," Joseph Madachy, Vol. 5, October 1972, pages 253-54.

"Consecutive-Digit Primes (Round 3)," Raphael Finkelstein and Judy Leybourne, Vol. 6, Summer 1973, pages 204-06.

"Consecutive-Digit Primes (Finale)," Ray P. Steiner (formerly Raphael Finkelstein), Vol. 10, No. 1, 1977-78, pages 30-31.

In March 1978 I published the remarkable discovery of Alan Cassel that the sequence 123456789, repeated seven times and followed by 1234567, is a pseudo prime with odds of about a trillion to one that it is a genuine prime. In December of the same year I was able to report that physicist R. E. Crandall and computer scientist Michael A. Penk proved that the 70-digit number is indeed prime. It is the largest such prime known.

Harry Nelson, in the *Journal of Recreational Mathematics* (Vol. 10, No. 1, 1977-78, page 33), raised an interesting question about the number formed by writing the counting numbers in sequence, 123456789101112131415 . . . . Taken as a decimal fraction, the number is known to be transcendental. If truncated at some spot, is the number ever prime? Nelson tested it through 2^{48} decimals without finding a prime. He conjectures that the question is undecidable.

Similar questions can, of course, be raised about the primality of the first n digits of any irrational number. In the case of pi, four such primes are known: 3, 31, 314159, and 3 14159 26535 89793 23846 26433 83279 50288 41. The last number was proved prime in 1979 by Robert Baillie and Marvin Wunderlich. Baillie has checked pi through 432 decimals without finding another prime. Is there a fifth? Probably, but it may be a long time before anyone knows.

Steward Hartman, assisted by Kenneth Bell, sent the most complete compilation of words spelled with the initial letters of the names of the ten digits. They range from five letters (for example "often") through twelve. Many readers hit on the eight-letter words "footnote," "fossette," and "nonsense," each of which can be lengthened to nine letters by making it plural. Eleven-letter words include "oftennesses," "tensenesses," and "sostenentes." ("Sostenente" means "sustained" in music.) Edward Kulkosky, Raymond J. Lovett, and Stephen C. Koehler were the first to hit the jackpot with the thirteen-letter "noneffeteness," which can be stretched to fifteen by adding "es." In the category of plausible made-up words, David Elwell proposed "nontennesseeness." Add "es" and the length is eighteen. Many readers pointed out that "Tennessee" qualifies as a proper name.

John Gribbin followed his book *The Jupiter Effect*, in which he predicted the destruction of Los Angeles in 1982, with an even funnier one, *The Jupiter Effect Reconsidered* (Vintage, 1982). For a review, see "The Gribbin Effect," reprinted as Chapter 39 of my *Order and Surprise*.

Gribbin's prediction was based on pseudoscience, but predictions of earthquakes by self-styled psychics have been so numerous that a listing would fill a volume. Eventually, of course, by sheer laws of chance, some psychic will hit the bull's eye with an accurate prediction that specifies the day and place of a major quake. When that happens, believers in precognition will naturally be enormously impressed, and the lucky psychic will become instantly famous. To date, unfortunately, the guesses of psychics have either been spectacular failures or so vague as to be worthless. This seems odd. If psychics can predict such things as murders and airplane crashes one would suppose that the sudden deaths of tens of thousands would send a much stronger information wave backward in time.

When I asked Dr. Matrix why his quake prediction had failed, he gave the following explanation. He had lied, he said, when he told me the quake would be small. Actually, his cockroaches had predicted a catastrophe that would reduce both Los Angeles and San Francisco to rubble and kill millions. He had not told me this, he insisted, because he feared the panic it would cause if I printed it.

Word of Dr. Matrix's prediction reached Jerry Brown, then governor of California, who at once summoned Dr. Matrix to his office. The governor had been impressed by my suggestion that the quake could be triggered by the combined PK powers of California cockroaches. Surely, he told Dr. Matrix, the psi powers of cockroaches must be very small compared to the powers of human psychics. Dr. Matrix agreed.

The two men got in touch with the U.S. government, and the CIA acted quickly. Of course their plan was then top secret, and I am revealing it now for the first time. Even Jack Anderson, who got wind of it from an informant, agreed to remain silent. The CIA provided transportation for bringing to the Stanford area the greatest psychics of the world. Nina Kulagina was flown from Moscow, Uri Geller from Mexico, Ingo Swann from New York, and Jean-Pierre Girard from Paris. Ted Serios could not be located, but a Japanese boy even more skilled at projecting his thoughts onto Polaroid film, was brought from Tokyo. For seven days the superpsychics concentrated their psi energy on the San Andreas fault. Slowly the mechanical stress that had been building up began to dissipate. The Palmdale Bulge subsided. Los Angeles and San Francisco were saved. Governor Brown also played a significant role in the plan by initiating a massive campaign to exterminate the state's cockroaches.

"Did you return any of the fees you collected from the clients to whom you sold your prediction?" I asked.

Dr. Matrix looked dumbfounded. "Certainly not! They paid for a prediction. They got a genuine one. You can no more blame me for the happy outcome than you can blame Jonah for his failed prophecy after Jehovah changed his mind about destroying Nineveh."

If any readers think the use of cockroaches to predict earthquakes is ridiculous, let me recommend the chapter on earthquake prediction in Jeffrey Goodman's book *We are the Earthquake Generation: When and Where the Catastrophes Will Strike* (Seaview Books, 1978). Dr. Goodman is best known as a psychic archeologist who uses psi powers for making archeological finds. "Someday a cockroach may save your life," he writes. He concedes that the little creatures may be responding to tiny tremors of some sort that precede a quake, but he thinks "ESP, rather than some physical sense, may be the most fruitful area of research on this topic." Here is his conclusion:

It is in precognition . . . that animals appear to be genuinely gifted. This is precisely where we need assistance in earthquake research. . . . Ironically, the investigation of the ability of animals to sense earthquakes, now scientifically in vogue, seems to bring us back full circle to psychics, and their ESP ability to predict quakes.

For Helmut Schmidt's classic study of cockroach PK, see his "PK Experiments with Animals as Subjects," in the *Journal of Parapsychology*, Vol. 34, 1970, and articles by Schmidt and others in later issues. Louisa E. Rhine has an unintentionally hilarious section on Schmidt's cockroach experiments in her book *Psi, What Is It?* (Harper & Row, 1975), Chapter 17.

The most important tests of animal PK, subsequent to Schmidt's, were done in Rhine's laboratory by his director Walter J. Levy (on whose name I play in my column). As everybody interested in parapsychology knows, Dr. Levy was caught flagrantly cheating, resigned in disgrace, and hasn't been heard from since.

Twenty-one

I. My answer was *inkstand*, but lots of readers found another common dictionary word — *prankster*. Others sent in archaic words (given in the Oxford English Dictionary) such as *clinkstone, pinkster, pinkstone, sinkstone,* and *stinkstone*.

II. *Crankshaft* was the word I had in mind, but correspondents came up with *bankshare, monkship, monkshood,* and *tankship*.

Some readers composed sentences with solution words. Jim Rector ended his letter with "No thankstoyou, I thinkstoomuch."

III. If the positive integers are divided into even and odd numbers, any pair in either set will add to an even number greater than 2 and thus cannot add to a prime.

IV. The social security number is 381-65-4729. Adding 0 at the end gives the unique solution to the same problem with the ten digits from 0 through 9, and the ten-digit number divisible by 10.

Scores of readers wrote to insist that 381654729 is not unique and to offer alternate solutions. I found the situation mystifying until I realized what had happened. They had all relied on a pocket calculator with an eight-digit readout to divide an eight-digit number by 8, without checking by hand to see if there was a remainder! (Incidentally, a large number is divisible by 8 if and only if its last three digits are.) Many other readers sent correct proofs of the uniqueness of the number 381654729, based on either computer programs or on familiar divisibility rules.

Michael R. Leuze worked with a computer to examine solutions to this problem in number systems other than base 10. He found that there are no solutions in any odd base or in base 12. In base 2 there is the trivial solution 1. In base 4 there are two solutions, 123 and 321; in base 6, 14325 and 54321; in base 8, 3254167, 5234761, and 5674321. In base 14, as in base 10, there is a unique solution: 9 12 3 10 5 4 7 6 11 8 1 2 13. Leuze conjectures that there are no solutions in higher bases.

David M. Sanger generalized in a different way by asking what the numbers are whose first n digits are divisible by n, with no other requirements. His computer program found all such numbers, starting with the 45 two-digit numbers (ignoring initial zeros) and ending with the unique 25-digit number 3,608,528,850,368,400,786,036,725.

Because no 26th digit can be added to make a number divisible by 26, that number ends the sequence. There are 2,492 10-digit numbers; the smallest is 1,020,005,640 and the largest is 9,876,545,640. One peculiar number is 3,000,060,000. The number of numbers increases steadily from $n = 2$ through $n = 9$ and $n = 10$ (in both cases there are 2,492 numbers) and then declines steadily. For $n = 20$ through $n = 25$ the numbers of numbers are respectively 44, 18, 12, 6, 3 and 1.

Allen Tarr, of Winnipeg, found a startling way in which the *lo shu*, the ancient Chinese magic square, is related to the number 381654729. Figure 44 is self-explanatory.

Warren Holland wrote to say that the number could not be a social security number because when the fourth and fifth digits of such a number are greater than 09, both digits must be even.

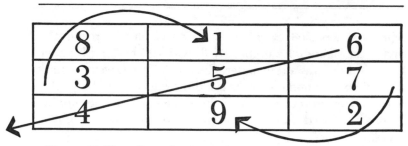

Figure 44. How the order-3 magic square generates 381 654 729

He said that this is one of the rules for identifying social security numbers, citing references which I was not able to check.

Leuze sent a thorough analysis of the problem in negative bases. For other generalizations, see "Progressively Divisible Numbers," by Stewart Metchette, in *Journal of Recreational Mathematics*, Vol. 15, No. 2 (1982-83), pages 119-22. The problem surfaced again in "What's In a Number?", *Mathematical Gazette*, Vol. 67 (December 1983), pages 281-282. The editor added in a footnote that when the problem appeared in London's *Sunday Times* in 1982, irate readers made the same careless mistake some of my readers did — they complained that the answer was not unique because they had worked on the problem with an eight-digit calculator!

Jaime Poniachik (the person who posed the problem in my column) told me the problem had been invented by his wife, Lea Gorodisky, who worked with him on the *Snark*. This magazine no longer exists, having been followed by the puzzle magazine *Juegos*. Poniachik is an editor. It is a delightful magazine, with very funny cartoons that often relate to games and recreational mathematics.

V. A 16-move solution to the knights-switch problem is given below. Rows are numbered 1 through 4 from bottom to top, and columns are labeled *A*, *B*, and *C* from left to right.

1. A1-C2	9. A2-C1
2. C2-A3	10. B1-C3
3. B4-C2	11. C3-A4
4. C2-A1	12. A3-C2
5. C1-A2	13. C4-A3
6. A2-B4	14. A3-B1
7. A4-C3	15. C2-A3
8. C3-A2	16. A3-C4

Alan Delahoy was the first of many readers who sent proofs that 16 moves are minimal for the knight-switch problem. Some of these proofs began by transforming the problem to a 12-point graph in the manner given in my book *Aha! Insight* (1977). When the problem is in this form, it is easy to show that a solution must have an even number of moves that cannot be less than 14. (If the board has only three knights at one end, seven moves are needed to get them from one end to the other.) All that remains, then, is to show that 14 moves are too few. The insight that reduces 18 moves to 16 is the realization that back-tracking one knight—returning it to a cell from which it had moved—clears the way for another knight move. All 16-move solutions have this feature.

Howard Rumsey, Jr., proved on a home computer that any distribution of the six knights can be reached from the starting pattern in 22 moves or fewer. Any knight pattern can be reached from any other pattern in no more than 26 moves. Rumsey found seven pairs of such patterns (not counting symmetries) for which 26 moves are required to go from one distribution of knights to the other.

James G. Mauldon suggested replacing the middle knight on each side with a king that moves like a knight, and adding the requirement that the kings change places. He found a 22-move solution. If knights are placed on the two central squares to make four knights of the same color on each side, the knights of different colors can be switched in 12 moves. Mauldon also added the proviso that these eight knights be paired with respect to the horizontal center line of the board and that all pairs be interchanged. He believes 36 moves are minimal here. He found a 44-move solution to the problem that pairs the knights that are symmetrically opposite with respect to the center of the board, but this was improved by Craig Collins to 40 moves.

If we add the proviso that the black and white knights must alternate moves, as in chess, we have the form of the problem given by Henry Ernest Dudeney in *The Canterbury Puzzles* (1907), Problem 94. He gives the best solution — 22 moves.

George Starbuck found "schnappsed," another ten-letter word of one syllable, but he was topped by his friend William Harmon (both men are poets), who wrote to him: "Schnappsed can't be beat. I realized this while being broughammed to the airport." (Several dictionaries prefer the monosyllabic pronunciation of this 11-letter word.

The robot's reasoning as to why a person cannot be exactly one-third Scottish, one-third Chinese and one-third Hungarian came in for heavy criticism. Readers were right in not taking the robot's solution seriously, since it relies on several unrealistic assumptions. In particular, ASMOF'S argument assumes that all progenitors are full-blooded. In the real world this becomes preposterous as soon as someone's ancestry is traced back a few generations.

Twenty-two

The four prices Iva paid for the trinkets at the Great Bazaar are $1, $1.50, $2, and $2.25. Both the sum and the product of this set are $6.75. If it had not been specified that Iva paid $1 for the earrings, there would have been a second solution to the problem: $1.20, $1.25, $1.80, and $2.50. The problem, posed by Kenneth M. Wilke (in *Crux Mathematicorum*, Vol. 4, June 1978), is a variant of one devised by David A. Grossman that appears as No. 65 in L. A. Graham's *Ingenious Mathematical Problems and Methods* (Dover Publications, 1959). In Grossman's version both the sum and the product are $7.11, and the unique set of prices is $1.20, $1.25, $1.50 and $3.16.

The simplest way to dissect a cube into three congruent solids is to slice it into three parallel slabs. When I gave this absurd answer, I foolishly said I knew of no other way to accomplish such a trisection. As many readers were quick to tell me, I could not have been more wrong.

John E. Morse sent the most general solution. If you hold a cube so one corner points toward you and the outline appears to be a regular hexagon, you will see the cube's 3-symmetry. This symmetry makes it possible to slice the cube into three congruent parts in an infinite number of ways. The surfaces of the parts may be flat or curved in any manner, and it is easy to design weird trisections for which the parts are so intricately interlocked that they cannot be separated.

Figure 45 shows how to label the corners of a cube with numbers 0 through 7 so that the sum of the pair of numbers at the ends of each edge is a prime. The solution is unique, not counting rotations and reflections. It is not hard to show there is no solution using eight consecutive integers that start with any number greater than zero. The problem was sent to me by its originator, Garry Goodman.

Bennett Battaile suggested a companion problem. Using the same eight digits (0 through 7), can each sum be a composite number? The answer is yes, and this too has a unique solution. I leave it as an unanswered problem for the interested reader. Leslie Card found a way to label a cube's corners with square

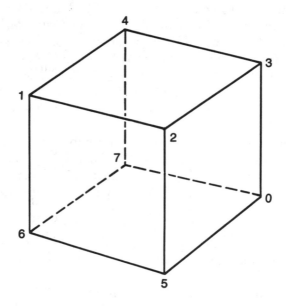

Figure 45. Solution to the cube-labeling problem

numbers, no two alike, so that each sum is a prime, but he did not try to prove the solution unique or to look for others.

On Dr. Matrix's hinged cube see *Mathematical Models*, by H. Martyn Cundy and A. P. Rollett (Oxford University Press, second edition, 1961, page 122) and letters in the *Mathematical Gazette* (Vol. 57, No. 399, pages 66-67, February 1973, and Vol. 57, No. 401, page 211, October 1973).

The paradox of the surface-to-volume ratios of the sphere and the cube is the first problem in Christopher P. Jargocki's *Science Brain-Twisters, Paradoxes and Fallacies* (Charles Scribner's Sons, 1976). The fallacy springs from the fact that two entirely different meanings are applied to the symbol d. The proper way to compare the surface-to-volume ratios of the sphere and the cube is to consider two solids of equal volume. A cube of volume v has an edge of $v^{1/3}$ and a surface-to-volume ratio of $6(v^{1/3})^2/v$ or $6/v^{1/3}$. A sphere of the same volume has a diameter of $(6v/\pi)^{1/3}$ and a surface-to-volume ratio of $\pi(6v/\pi)^{2/3}/v$ or about $4.8/v^{1/3}$. The surface area of a sphere, Jargocki points out, is about 20 percent less than that of a cube of the same volume. Similar reasoning shows that for a given surface area the volume of a sphere exceeds that of a cube by about 39 percent.

The following surprising theorem, sent to me by Thomas D. Waugh, is a closely related fallacy. If a sphere of radius r is enclosed in a polyhedron in which each face touches the sphere, the surface-to-volume ratio of the polyhedron, regardless of its shape or number of faces, is $3/r$. But $3/r$ is also the surface-to-volume ratio of the sphere! The two paradoxes are good introductions to what physicists call dimensional analysis, the technique of expressing constants in dimensionless numbers independent of adopted units of measurement.

As Dr. Matrix pointed out, no number with more than ten digits can be a no-rep emirp because it forces a duplication of digits. Harvey P. Dale and others wrote to say that not even a ten-digit number can be a no-rep emirp because any permutation of the ten digits will have a sum of 45, or a digital root of 9. This means it will be a multiple of 9, and therefore composite. The highest no-rep emirp is 987653201.

I asked if there is a cyclic emirp with more than seven digits. Joseph Buhler has shown that there is no such emirp with eight, nine, or ten digits. Thus aside from the trivial two-digit cyclic emirps, the only known cyclic emirp is the six-digit 193939. It may be difficult to prove there are no larger ones.

Dr. Matrix mentioned the unsolved problem of determining the last prime on an alphabetical listing of the English names of primes. (See "Alphabetizing the Integers," by Edward R. Wolpow, in *Word Ways*, Vol. 13, February 1980.) Donald E. Knuth, the noted computer scientist at Stanford University, gave the task to his students in a programming seminar. More than half a dozen students independently found the answer: two vigintillion two undecillion two trillion two thousand two hundred ninety-three.

A number of readers caught the joke in Dr. Matrix's remark that 19 is an unusual prime because 19 is the sum of 9 and 10, and also the difference between the squares of 9 and 10. *Every* odd number can be partitioned into two consecutive integers that add to the number, and have the same number as the difference between their squares.

Dr. Khalifa is no joke. He is a devout Muslim, passionately convinced that the Koran is the inspired word of Allah. When he wrote to me in 1980 he was senior chemist in the State Chemist's Office of Arizona, in Mesa. After graduating at Ain Shams University in Cairo, he obtained a master's in chemistry at the University of Arizona, and in 1964 received his doctorate at the University of California, Riverside. He has published some two dozen scientific papers.

The pamphlet Dr. Matrix showed me was out of print in 1980. It contained, Khalifa said, only a fraction of the marvelous numerological discoveries he has made. He sent me an unpublished paper titled "The Existence of God: Finally, Scientific

Proof." It is his conviction that the "overwhelming pervasive-
ness" (as he put it) of 19 in the Koran can be explained only
by assuming the book's divine origin.

Many readers informed me that both the words and music
of ballad from which I took my epigraph were written in the
1870s by William Percy French (1854-1920), an Irish music-
hall entertainer. A London publisher printed the song
without acknowledgment, there were many other pirated
editions, and to this day its author is usually listed in antholo-
gies as "anonymous." Soldiers in the First World War liked to
belt it out, often with improvised bawdy stanzas. A Victor
recording, sung by Frank Crumit, appeared in the twenties.

No two printings of the lyrics seem to be alike. Two
markedly different versions, together with the music, can be
found in Sigmund Spaeth, *Read 'em and Weep* — *The
Songs You Forgot to Remember* (1926), and Carl Sandburg,
The American Songbag (1927). Apparently neither compiler
knew that the ballad had an Irish origin. Percy French's
original version—it begins "Oh the sons of the prophets are
hardy and grim . . ." — can be found in *The Best of Percy
French,* a booklet published in 1980 by EMI Publishing,
Ltd., 138-140 Charing Cross Road, London.

So that readers may appreciate the remarkable way in
which Dr. Matrix's duel parallels the tragic action of the song,
I reprint here (with a few small changes) the words as they
are given in Ralph Wood, *A Treasury of the Familiar* (1948).

The Sons of the prophet are brave men and bold,
And quite unaccustomed to fear,—
But the bravest by far in the ranks of the Shah
Was Abdul A-bul-bul A-mir. [1]

1. In some versions of the ballad it is Abdul the Bulbul Amir. Bulbul, I am
told, is Arabic for nightingale.

If you wanted a man to encourage the van
Or harass the foe from the rear,
Storm fort or redoubt, you had only to shout
For Abdul A-bul-bul A-mir.

Now the heroes were plenty and well known to fame
In the troops that were led by the Czar
And the bravest of these was a man by the name
Of Ivan Skavinsky Skavar.

He could imitate Irving,[2] play poker and pool,
And strum on the Spanish guitar,
In fact quite the cream of the Muscovite team
Was Ivan Skavinsky Skavar.

One day this bold Russian had shouldered his gun,
And donned his most truculent sneer,
Downtown he did go, where he trod on the toe
Of Abdul A-bul-bul A-mir.

"Young man," quoth Abdul, "has your life grown so dull
That you wish thus to end your career?
Vile infidel, know, you have trod on the toe
Of Abdul A-bul-bul A-mir."

Said Ivan, "My friend, your remarks in the end
Will avail you but little, I fear,
For you ne'er will survive to repeat them alive,
Mr. Abdul A-bul-bul A-mir."

2. Not Irving Joshua Matrix, but Sir Henry Irving, a famous Victorian actor
   who often played opposite Ellen Terry.

"So take your last look at the sunshine and brook,
And send your regrets to the Czar—
For by this I imply, you are going to die,
Count Ivan Skavinsky Skavar!"

Then this bold Mameluke[3] drew his trusty skibouk,[4]
With a cry of "Allah Akbar,"
And with murderous intent he ferociously went
For Ivan Skavinsky Skavar.

They parried and thrust, and they sidestepped and cussed,
Of their blood they spilled a great part;
The philologist blokes, who seldom crack jokes,
Say that hash was first made on that spot.

They fought all that night, 'neath the pale yellow moon,
The din, it was heard from afar,
And huge multitudes came, so great was the fame,
Of Abdul and Ivan Skavar.

As Abdul's long knife was extracting the life,
In fact he was shouting "Huzzah,"
He felt himself struck by that wily Kalmuck,[5]
Count Ivan Skavinsky Skavar.

3. According to the Oxford English Dictionary a mamaluke is "a member
   of the military body, originally composed of Caucasian slaves, which
   seized the throne of Egypt in 1254, and continued to form the ruling
   class in that country until the early part of the 19th century."
4. Skibouk: I assume this is a dagger.
5. The Kalmucks were Asiatic Buddhists who settled in the lower Volga
   basin of Russia. The Kalmuck region of the USSR was dissolved because of Kalmuck collaboration with the German invaders of World
   War II, but reestablished in 1958.

The Sultan drove by in his red-breasted fly,[6]
Expecting the victor to cheer,
But he only drew nigh to hear the last sigh
Of Abdul A-bul-bul A-mir.

Czar Petrovitch too, in his spectacles blue,
Rode up in his new-crested car.
He arrived just in time to exchange a last line,
With Ivan Skavinsky Skavar.

There's a tomb rises up where the Blue Danube rolls,
And 'graved there in characters clear,
Are, "Stranger, when passing, oh pray for the soul
Of Abdul A-bul-bul A-mir."

A splash in the Black Sea one dark moonless night,
Caused ripples to spread wide and far,
It was made by a sack fitting close to the back
Of Ivan Skavinsky Skavar.

A Muscovite maiden her love vigil keeps,
'Neath the light of a pale polar star,
And the name that she murmurs so oft as she weeps
Is Ivan Skavinsky Skavar.

6. Fly: A fast-moving small carriage usually drawn by a simgle horse.